BARRON'S
Painless
Geometry

Fourth Edition

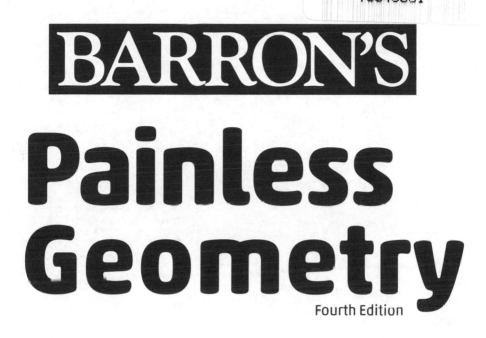

Lynette Long, Ph.D.

Published by Kaplan, Inc., d/b/a Barron's Educational Series
750 Third Avenue
New York, NY 10017
www.barronseduc.com

ISBN: 978-1-5062-6804-0

10 9 8 7 6 5 4 3 2

Kaplan, Inc., d/b/a Barron's Educational Series print books are available at special quantity discounts to use for sales promotions, employee premiums, or educational purposes. For more information or to purchase books, please call the Simon & Schuster special sales department at 866-506-1949.

Contents

How to Use This Book

Painless geometry? Impossible, you think. Not really. I have been teaching math or teaching teachers how to teach math for more than twenty years. Math is easy . . . or at least it can be with the help of this book!

Whether you are learning geometry for the first time, or you are trying to remember what you've learned but have forgotten, this book is for you. It provides a clear introduction to geometry that is both fun and instructive. Don't be afraid. Dive in—it's painless!

Painless Icons and Features

This book is designed with several unique features to help make learning geometry easy.

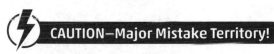

You will see Painless Tips throughout the book. These include helpful tips on the topics in that section.

Caution boxes will help you avoid common pitfalls or mistakes. Be sure to read them carefully.

BRAIN TICKLERS

There are Brain Ticklers throughout each chapter in the book. These quizzes are designed to make sure you understand what you've just learned and to test your progress as you move forward in the chapter. Complete all the Brain Ticklers and check your answers. If you get any wrong, make sure to go back and review the topics associated with the questions you missed.

SUPER BRAIN TICKLERS

Super Brain Ticklers appear at the end of each chapter. These are comprehensive quizzes on all of the topics covered in the chapter. Solve these questions to make sure that you understand all of the chapter topics before moving forward.

PAINLESS STEPS

Complex procedures are divided into a series of painless steps. These steps help you solve problems in a systematic way. Follow the steps carefully, and you'll be able to solve most geometry problems.

EXAMPLES

Most problems are illustrated with examples. Study each of the examples. If you are having trouble, research shows that writing/copying the problem and the solution may help you understand it.

ILLUSTRATIONS

Painless Geometry is full of illustrations to help you better understand geometry topics. Geometry is a visual subject. If you can visualize a problem, it will be easier to understand it.

MINI-PROOFS

Numerous mini-proofs appear throughout the book. Written informally, these mini-proofs are the first step toward understanding formal geometry proofs.

EXPERIMENTS

The book contains a series of experiments designed to help deepen your understanding of geometry. Why not try some of the experiments? They are fun and informative.

Geometry is a linear subject, and what you don't know in one chapter will haunt you in the next. Don't move forward if any topic confuses you. Go back and practice those topics until they are clear.

Chapter Breakdown

Chapter One is titled "A Painless Beginning," and it really is. It is an introduction to the basic terms and symbols of geometry.

Chapter Two, "Angles," will teach you everything about angles. It will show you how to measure and classify different types of angles as well as explain the relationship between different angle pairs.

Chapter Three, "Parallel and Perpendicular Lines," will show you how to identify parallel and perpendicular lines. You will learn what types of angles are formed when parallel and perpendicular lines are formed.

Chapter Four, "Triangles," shows you that there is more to learn about triangles than you ever dreamed possible. You will learn the types of triangles, as well as how to find the perimeter and area of a triangle.

Chapter Five, "Similar and Congruent Triangles," explains what similar and congruent triangles are. By the time you finish reading this chapter, you will know what SSS, SAS, ASA, and AAS mean and how to use each of them to prove that two triangles are congruent.

Chapter Six is titled "Quadrilaterals," which is a fancy word for four-sided figures. Trapezoids, parallelograms, rhombuses, rectangles, and squares are all quadrilaterals. You will learn all about these interesting polygons here.

Chapter Seven is titled "Polygons," and you will learn that polygons are everywhere. The frame of your television is a polygon, and so is the STOP sign down the street. You will be able to impress your friends when you know the names of your polygons, and you'll learn the fascinating things all regular polygons have in common.

Chapter Eight could have been titled "Circles, Circles, and More Circles." You may know what a circle is, but you will find out how a pie is not only something you eat but an essential term in geometry.

Chapter Nine, "Perimeter, Area, and Volume," will show you how to find the perimeter, area, and volume of some common and not so common shapes. It's painless, just watch.

Chapter Ten, "Graphing," will teach you how to graph a point and a line. You can use what you learn here to draw your own geometric shapes.

Chapter Eleven, "Constructions," will teach you how to construct or copy shapes with just a compass, straight edge, and pencil. When you're ready for a break, use these same tools to create some interesting designs.

Chapter 1

A Painless Beginning

Geometry is a mathematical subject, but it is just like a foreign language. In geometry, you have to learn a whole new way of looking at the world. There are hundreds of terms in geometry that you've probably never heard before. Geometry also uses many common terms, but they have different meanings. A point is no longer the point of a pencil, and a plane is not something that flies in the sky. To master geometry, you have to master these familiar and not so familiar terms, as well as the theorems and postulates that are the building blocks of geometry. You have a big job in front of you, but if you follow the step-by-step approach in this book and do all the experiments, it can be painless.

Undefined Terms

Undefined terms are terms that are so basic they cannot be defined, but they can be described.

A *point* is an undefined term. A point is a specific place in space. A point has no dimension. It has no length, width, or depth. A point is represented by placing a dot on a piece of paper. A point is labeled by a single letter.

A *line* is a set of points that extend indefinitely in either direction. In this book, the term *line* will always mean a straight line.

This line is called line AB, or line AC, or line BC, or $\overleftrightarrow{AB}$, $\overleftrightarrow{AC}$, or $\overleftrightarrow{BC}$. The two-ended arrow over the letters tells you that the letters are describing a line. You can name a line by any two points that lie on the line.

A *plane* is a third undefined term. A plane is a flat surface that extends infinitely in all directions. If you were to imagine a tabletop that went in every direction forever, you would have a plane.

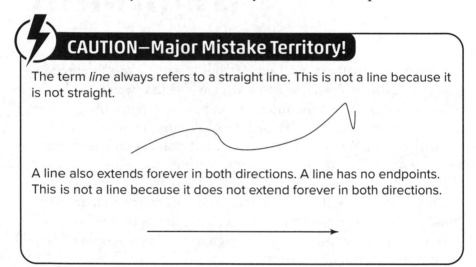

CAUTION—Major Mistake Territory!

The term *line* always refers to a straight line. This is not a line because it is not straight.

A line also extends forever in both directions. A line has no endpoints. This is not a line because it does not extend forever in both directions.

A plane is often represented by a four-sided figure. Place a capital letter in one of the corners of the figure to name the plane. This is plane P.

P

Some Defined Terms

A line segment is part of a line with two endpoints. A line segment is labeled by its endpoints. We use two letters with a bar over them to indicate a line segment. This line segment is segment XY or $\overline{XY}$.

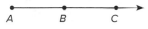

A *ray* is a part of a line that has one endpoint and extends infinitely in the other direction. The endpoint and one other point on the ray label a ray. This ray could be labeled ray AB or ray AC. When you label a ray, put a small arrow on the top of the two letters to indicate that it is a ray. The arrow on top of the two letters should show the direction of the ray. This ray could also be labeled as $\overrightarrow{AB}$ or $\overrightarrow{AC}$. This is *not* $\overrightarrow{BC}$ since B is not an endpoint.

Opposite rays are two rays that have the same endpoint and form a straight line. Ray BA and ray BC are opposite rays. We can also say that $\overrightarrow{BA}$ and $\overrightarrow{BC}$ are opposite rays.

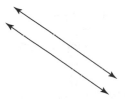

Parallel lines are two lines in the same plane that do not intersect.

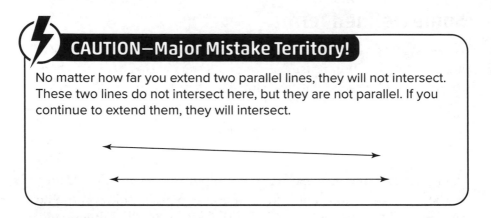

CAUTION—Major Mistake Territory!

No matter how far you extend two parallel lines, they will not intersect. These two lines do not intersect here, but they are not parallel. If you continue to extend them, they will intersect.

Collinear points lie on the same line. *A*, *B*, and *C* are collinear points. *D* is not collinear with *A*, *B*, and *C*.

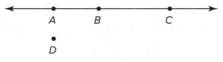

Noncollinear points do not lie on the same line. *X*, *Y*, and *Z* are noncollinear points. All three of them cannot lie on the same line.

BRAIN TICKLERS   Set # 1

Decide whether each of the following statements is true or false.

1. A ray has one endpoint.

2. Points that lie on the same line are called linear.

3. A line is always straight.

4. A line segment can be ten miles long.

(Answers are on page 14.)

More Defined Terms

An *angle* is a pair of rays that have the same endpoint.

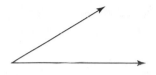

The rays are the *sides* of the angles. The endpoint where the two rays meet is called the *vertex* of the angle.

A *polygon* is a closed figure with three or more sides that intersect only at their endpoints. The sides of the polygons are line segments. The points where the sides of a polygon intersect are called the vertices of the polygons.

These are all polygons.

Two shapes are *congruent* if they have exactly the same size and shape.

Each of these pairs of figures is congruent.

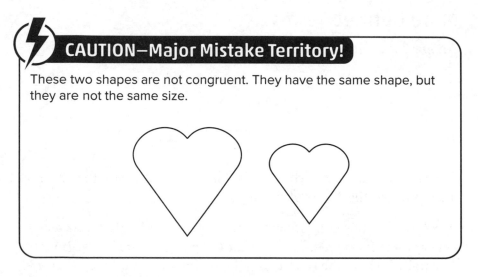

CAUTION—Major Mistake Territory!

These two shapes are not congruent. They have the same shape, but they are not the same size.

The *perimeter of a polygon* is the sum of the lengths of all the sides of the polygon. The perimeter of a polygon is expressed in linear units such as inches, feet, meters, miles, and centimeters.

The *area* of a geometric figure is the number of square units the figure contains. The area of a figure is written in square units such as square inches, square feet, square miles, square centimeters, square meters, and square kilometers.

The *ratio* of two numbers a and b is a divided by b, written $\frac{a}{b}$ as long as b is not equal to zero.

A *proportion* is an equation that sets two ratios equal. $\frac{a}{b} = \frac{c}{d}$ is an example of a proportion.

BRAIN TICKLERS   Set # 2

Determine whether each of the following statements is true or false.

1. A polygon can be composed of three segments.

2. The perimeter of a figure can be measured in square inches.

3. 5/0 is a ratio.

(Answers are on page 14.)

Postulates

Postulates are generalizations in geometry that cannot be proven true. They are just accepted as true. Do the following experiment to discover one of the most basic postulates of geometry.

EXPERIMENT

Find out how two points determine a line.

Materials
Pencil
Paper
Ruler

Procedure
1. Draw two points on a piece of paper. To draw a point make a dot on the paper as if you were dotting the letter i.

2. Draw a line through these two points. Now draw another line through these same two points. Draw a third line through these two points. Are the three lines the same or different?

3. Draw two other points. How many different lines can you draw through these two points?

Something to think about . . .
How many lines can you draw through a single point?

POSTULATE   Two points determine a single line.

What does this postulate mean?

- You can draw many lines through a single point.
- You can only draw one line through any two different points.

POSTULATE   At least three points not on the same line are needed to determine a single plane.

What does this postulate mean?

- Through a single point there are an infinite number of planes.
- Through two points you can also find an infinite number of planes.

- Through three points that lie in a straight line there are an infinite number of planes.
- Through three points that do not lie in a straight line there is only one plane.

POSTULATE   If two different planes intersect, they intersect on exactly one line.

What does this postulate mean?

- Not all planes intersect.
- Two different planes cannot intersect on more than one line.

Theorems

Theorems are generalizations in geometry that can be proven true.

THEOREM   If two lines intersect, they intersect at exactly one point.

What does this theorem mean?

- Not all lines intersect.
- Two different lines cannot intersect at more than one point.

This statement is a theorem and not a postulate because mathematicians can prove this statement. They do not have to accept it as true.

THEOREM   Through a line and a point not on that line, there is only one plane.

Three noncollinear points determine a single plane. A line provides two points and the point not on that line is the third noncollinear point.

Conditional Statements

Conditional statements are statements that have the form "If p, then q."

An example of a conditional statement is "If a figure is a square, then it has four sides."

The first part of the conditional statement is called the *hypothesis*. In the above example, "If a figure is a square" is the hypothesis. The second part of the conditional statement is called the *conclusion*. In the above example, "then it has four sides" is the conclusion. Many theorems are conditional statements. If the first part of the statement is true, then the second part of the statement has to be true.

The *converse* of a conditional statement is formed by reversing the hypothesis and the conclusion. The converse of the conditional statement "If a figure is a square, then it has four sides" is "If a figure has four sides, then it is a square." The original statement is true, but the converse is not true. A rectangle has four sides, but it is not a square.

> **PAINLESS TIP**
>
> Write each postulate or theorem on a separate index card. On the back of the index card, draw a diagram to illustrate the postulate or theorem. Memorize all of them!

Geometric Proofs

In geometry, there are two types of proofs—indirect proofs and deductive proofs.

Indirect proofs

In an *indirect proof*, you assume the opposite of what you want to prove true. This assumption leads to an impossible conclusion, so your original assumption must be wrong.

Example:

You want to prove three is an odd number.
Assume the opposite. Assume three is an even number.
All even numbers are divisible by two.
Three is not divisible by two, so it can't be an even number.
Natural numbers are either even or odd, so three must be odd.

Deductive proofs

Deductive proofs are the classic geometric proofs. Deductive reasoning uses definitions, theorems, and postulates to prove a new theorem true. A typical deductive proof uses a two-column format. The statements are on the left and the reasons are on the right.

Statement	Reason

In this book, deductive proofs will have a different format. Three questions will be used to construct the proof. These three questions will help you think about proofs in a painless way.

1. What do you know?

2. What can you infer based on what you know?

3. What can you conclude?

Geometric Symbols

Much of geometry is written in symbols. You have to understand these symbols if you are going to understand geometry. Here are some common geometric symbols.

$=$ equal to
$3 + 5 = 8$
Three plus five is *equal to* eight.

$\neq$ not equal to
$6 \neq 2 + 1$
Six is *not equal to* two plus one.

$>$ greater than
$7 > 5$
Seven is *greater than* five.

$\geq$ greater than or equal to
$4 \geq 4$
Four is *greater than or equal to* four.

$<$ less than
$6 < 10$
Six is *less than* ten.

$\leq$ less than or equal to
$1 \leq 1$
One is *less than or equal to* one.

$\triangle$ triangle
$\triangle ABC$
Triangle ABC

$\cong$ congruent to
$\triangle ABC \cong \triangle DEF$
Triangle ABC is *congruent to* triangle DEF.

$\perp$ perpendicular to
line $AB \perp$ line CD
$\overleftrightarrow{AB} \perp \overleftrightarrow{CD}$
Line AB is *perpendicular to* line CD.

segment $XY \perp$ segment YZ

$\overline{XY} \perp \overline{YZ}$

Segment XY is *perpendicular to* segment YZ.

ray $AB \perp$ ray CD

$\overrightarrow{AB} \perp \overrightarrow{CD}$

Ray AB is *perpendicular to* ray CD.

π pi is a Greek symbol that represents the relationship between the ratio of the circumference of a circle to the diameter of a circle.
The value of π is constant.
It is estimated as 3.14.
The circumference of a circle is *pi* times the diameter of that circle.

$\sim$ similar to
$\triangle ABC \sim \triangle DEF$
Triangle ABC is *similar to* triangle DEF.

$\longleftrightarrow$ line
$\overleftrightarrow{AB}$
Line AB

$\longrightarrow$ ray
$\overrightarrow{AB}$
Ray AB

——— segment
$\overline{AB}$
Segment AB

⌒ arc
$\overarc{AB}$
Arc AB

$\parallel$ parallel to
$WX \parallel YZ$
Line WX is *parallel to* line YZ.

$\angle$ angle
$\angle A$
Angle A

BRAIN TICKLERS   Set # 3

Name the following symbols.

1. ≅

2. <

3. ⊥

4. ≥

(Answers are on page 14.)

> **PAINLESS TIP**
>
> Write each of the basic geometry symbols on the front of an index card. Write the word or words each symbol represents on the back. Memorize each of the symbols using the cards for help.

SUPER BRAIN TICKLERS

Match these definitions or symbols to the correct terms.

___ 1. ≥

___ 2. A specific place in space

___ 3. The number of square units inside a shape

___ 4. Part of a line with two endpoints

___ 5. ≅

___ 6. ≠

___ 7. Three points that lie on the same line

___ 8. Two rays with the same endpoint that form a line

A. congruent

B. collinear

C. greater than or equal to

D. point

E. area

F. segment

G. opposite rays

H. not equal to

I. equal to

(Answers are on page 14.)

BRAIN TICKLERS—THE ANSWERS

Set # 1, page 4

1. True
2. False
3. True
4. True

Set # 2, page 6

1. True
2. False
3. False

Set # 3, page 13

1. Congruent to
2. Less than
3. Perpendicular to
4. Greater than or equal to

Super Brain Ticklers, page 13

1. C
2. D
3. E
4. F
5. A
6. H
7. B
8. G

Chapter 2

Angles

An *angle* is two *rays* with the same endpoint. The common endpoint is the vertex of the angle. The rays are the sides of the angles. These two rays have the same endpoint.

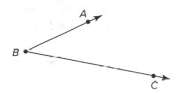

The endpoint B is the vertex of the angle. The sides of the angle are rays BA and BC.

Label an angle using three letters, one letter from one side, one letter from the other side, and the letter of the vertex. Remember that the letter of the vertex is in the middle. Label the above angle $\angle ABC$ or $\angle CBA$.

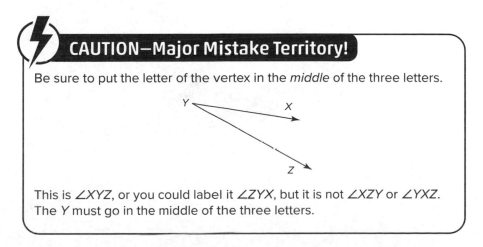

⚡ CAUTION—Major Mistake Territory!

Be sure to put the letter of the vertex in the *middle* of the three letters.

This is $\angle XYZ$, or you could label it $\angle ZYX$, but it is not $\angle XZY$ or $\angle YXZ$. The Y must go in the middle of the three letters.

An angle can also be labeled using the letter of the vertex. This angle could just be labeled ∠Y. You can only use one letter to label an angle if no other angle shares the same vertex.

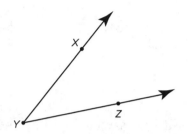

Perhaps the easiest way to label an angle is to put a number inside its vertex. When you use a number label, make a small arc inside the angle to make sure there is no confusion.

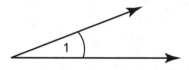

You can label this angle four different ways.

∠WXY
∠YXW
∠X
∠2

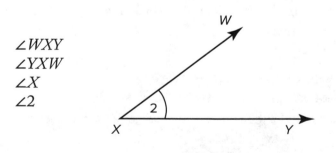

CAUTION—Major Mistake Territory!

If two angles share the same vertex, you cannot label them using a single letter.

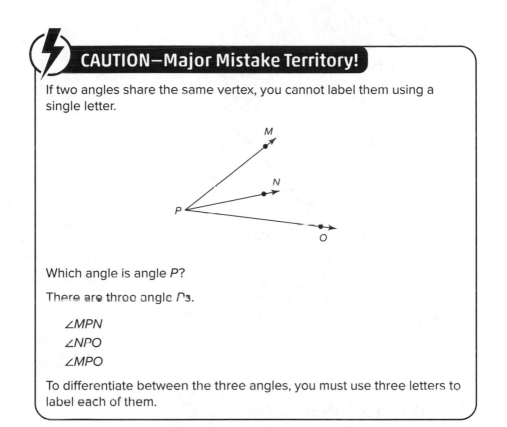

Which angle is angle *P*?

There are three angle *P*s.

∠MPN
∠NPO
∠MPO

To differentiate between the three angles, you must use three letters to label each of them.

An angle divides a plane into three distinct regions:

- interior of the angle
- angle itself
- exterior of the angle

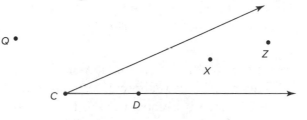

Points *X* and *Z* lie in the interior of the angle.
Points *C* and *D* lie on the angle.
Point *Q* lies on the exterior of the angle.

BRAIN TICKLERS   Set # 4

Based on the drawing, determine whether each of the following statements is true or false.

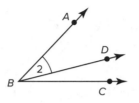

1. Each of the angles can be labeled in more than one way.

2. ∠ABD is the same as ∠DBA.

3. The measure of angle 2 is equal to the measure of ∠ABD.

4. There are two different angles that could be labeled ∠B.

(Answers are on page 34.)

Measuring Angles

You measure the length of a line segment using a ruler. You measure the size of an angle using a *protractor*. The unit of measure for an angle is the *degree*. A protractor is a semicircle that is marked off in 180 increments. Each increment is one degree.

PROTRACTOR POSTULATE   For every angle there corresponds a number between 0 and 180. The number is called the measure of an angle.

When expressing the measure of an angle, write "the measure of angle *ABC* is 25 degrees." This is abbreviated as m∠*ABC* = 25. The symbol for degrees is °, but it is generally omitted when measurements are written in the abbreviated format. However, if the measurement is given in the angle itself, the degree sign is used.

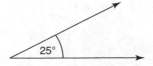

Using a protractor

Notice four things about a protractor.

1. There are numbers from 0 to 180.

2. The numbers are marked from right to left and from left to right around the curved part of the semicircle.

3. If the protractor is made of clear plastic, there is a horizontal line along the base of the protractor. The clear plastic line intersects the protractor at 0 and 180 degrees.

4. There is a mark at the halfway point of the straight side of a protractor.

To use a protractor, follow these painless steps:

Step 1: Place the center marker over the vertex of the angle.

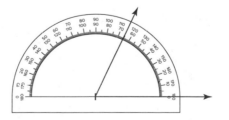

Step 2: Place the horizontal line along the base of the protractor on one of the sides of the angle.

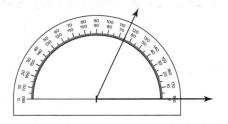

Step 3: Read the number where the other side of the angle intersects the protractor. This is the measure of the angle.

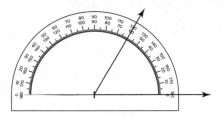

Watch as these angles are measured.

1. What is the measure of this angle?

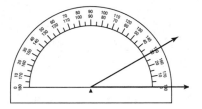

It measures 30 degrees.

2. Place the protractor on this angle.

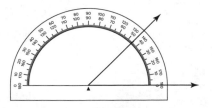

Notice that the angle measures 45 degrees.

3. Look at this angle. It measures 120 degrees.

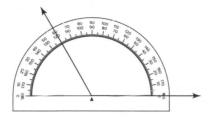

CAUTION—Major Mistake Territory!

When you measure angles, you are measuring the number of degrees in the angle, not the length of the sides. The length of the sides does not matter. Look at these two angles. Which is larger? Angle *A* or angle *B*?

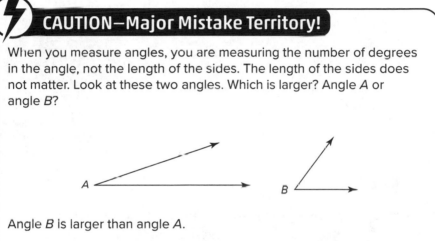

Angle *B* is larger than angle *A*.

Angle *A* has longer sides, but angle *B* contains more degrees.

BRAIN TICKLERS   Set # 5

Measure the following angles.

1.

2.

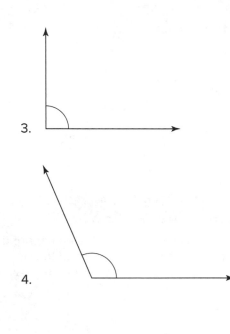

3.

4.

(Answers are on page 34.)

Angle Addition

You can add the measures of two angles together by adding the number of degrees in each angle together. For example, if the m∠A = 30 degrees and the m∠B = 20 degrees, then the m∠A plus the m∠B = 50 degrees.

You can also determine the difference between the size of two angles by subtracting the measure of the smaller angle from the measure of the larger angle.

If the m∠D = 45 degrees and the m∠E = 60 degrees, then the m∠E is 15 degrees larger than the m∠D, since 60 −45 = 15.

If two angles share the same vertex and a single side, their measures cannot be added if one angle is a subset of the other. For example, in the below illustration, the m∠ABD plus the m∠DBC is equal to the m∠ABC.

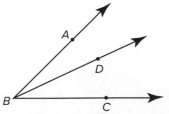

However, you cannot add the m∠*ABD* to the m∠*ABC*, since angle *ABD* is a subset of angle *ABC*.

BRAIN TICKLERS   Set # 6

The m∠*ABC* = 45 degrees and the m∠*CBD* = 50 degrees.

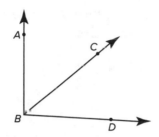

1. What is the sum of the two angles?

2. What is the difference between the two angles?

3. What is the difference between the m∠*ABC* and the m∠*ABD*?

(Answers are on page 34.)

Classifying Angles

Once you know how to measure angles you can classify them.

An *acute angle* measures less than 90 degrees and more than 0 degrees. These are all acute angles.

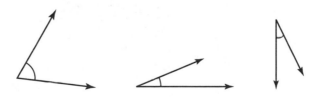

A *right angle* measures exactly 90 degrees. These are both right angles.

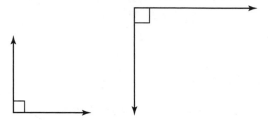

An *obtuse angle* measures more than 90 degrees but less than 180 degrees. These are all obtuse angles.

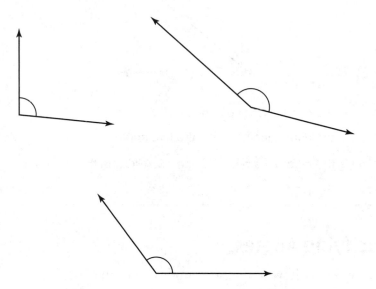

A *straight angle* measures exactly 180 degrees. This is a straight angle.

BRAIN TICKLERS   Set # 7

Decide whether each of the following statements is true or false.

1. A 63-degree angle is an acute angle.

2. A 90-degree angle is a straight angle.

3. A 179-degree angle is an obtuse angle.

4. A 240-degree angle is an obtuse angle.

(Answers are on page 34.)

Angle Pairs

If two angles have a total measure of 90 degrees, they are called *complementary angles*.

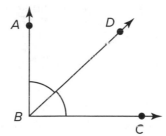

Angles *ABD* and *DBC* are complementary.

If the measure of angle *ABD* is 40 degrees, then the measure of angle *DBC* is 50 degrees.

$$40 + 50 = 90$$

To find the complement of an angle, use these two painless steps:

Step 1: Find the measure of the first angle.

Step 2: Subtract the measure of the first angle from 90 degrees.

Example:

Find the complement of a 10-degree angle.

Step 1: Find the measure of the first angle. It is given as 10 degrees.

Step 2: Subtract the measure of the first angle from 90 degrees.

$$90 - 10 = 80$$

The complement of a 10-degree angle is an 80-degree angle.

THEOREM   If two angles are complementary to the same angle, the measures of the angles are equal to each other.

Example:

If $m\angle A = 25$
and $\angle B$ is complementary to $\angle A$
and $\angle C$ is complementary to $\angle A$,
then $m\angle B = m\angle C$.
Because $\angle B$ is complementary to $\angle A$, $m\angle B$ is $(90 - 25)$ or 65.
Because $\angle C$ is complementary to $\angle A$, $m\angle C$ is $(90 - 25)$ or 65.

If two angles have a total measure of 180 degrees they are called *supplementary angles*.

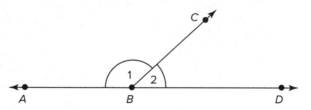

Example:

Angles ABC and CBD are supplementary.
If $\angle ABC$ measures 125 degrees, $\angle CBD$ must measure 55 degrees, since $125 + 55 = 180$.

To find the supplement of an angle, use these two painless steps:

Step 1: Find the measure of the first angle.

Step 2: Subtract the measure of the first angle from 180 degrees.

Example:

Find the supplement of a 10-degree angle.

Step 1: Find the measure of the first angle. It is given as 10 degrees.

Step 2: Subtract the measure of the first angle from 180 degrees.

$$180 - 10 = 170$$

The supplement of a 10 degree angle is a 170-degree angle.

THEOREM   If two angles form a straight line, the angles are supplementary.

Example:

Angle *ABD* and angle *DBC* form a straight angle. Angle *ABD* and angle *DBC* are supplementary to each other.

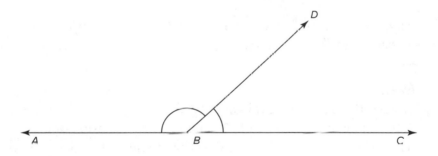

THEOREM   If two angles are supplementary to the same angle, the measures of the angles are equal to each other.

Example:

If m∠*A* = 70
and ∠*B* is supplementary to ∠*A*
and ∠*C* is supplementary to ∠*A*,
then m∠*B* = m∠*C*.
Because ∠*B* is supplementary to ∠*A*, m∠*B* is (180 – 70) or 110.
Because ∠*C* is supplementary to ∠*A*, m∠*C* is (180 – 70) or 110.

An *angle bisector* is a line or ray that divides an angle into two equal angles.

Example:

Ray *BD* bisects angle *ABC*. If angle *ABC* measures 60 degrees, then angle *ABD* measures 30 degrees, and angle *DBC* also measures 30 degrees.

$$30 + 30 = 60$$

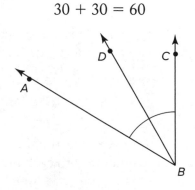

Four angles are formed when two lines intersect. The opposite angles always have the same measure, and they are called *vertical angles*.

Example:

Angles 1 and 3 are vertical angles. Angles 2 and 4 are also vertical angles. The measure of angle 1 is equal to the measure of angle 3. The measure of angle 2 is equal to the measure of angle 4.

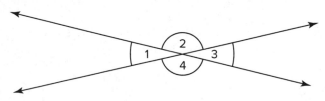

If angle 1 measures 70 degrees, then angle 3 must also measure 70 degrees.
If angle 2 measures 110 degrees, then angle 4 must also measure 110 degrees.

Two angles are *adjacent* if they share a common side and a common vertex but do not share any interior points.

Example:

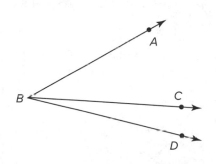

Angle *ABC* is adjacent to angle *CBD*. They share side *BC*.
Angle *ABD* is not adjacent to angle *CBD*. They share side *BD*, but they have common interior points.

⚡ **CAUTION—Major Mistake Territory!**

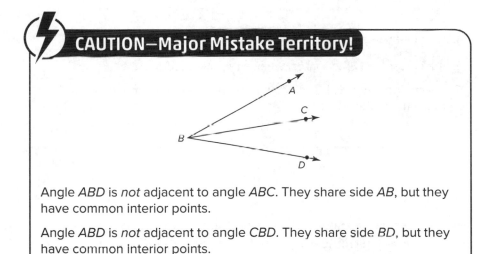

Angle *ABD* is *not* adjacent to angle *ABC*. They share side *AB*, but they have common interior points.

Angle *ABD* is *not* adjacent to angle *CBD*. They share side *BD*, but they have common interior points.

BRAIN TICKLERS   Set # 8

1. If ∠A measures 59 degrees, what is the measure of its complement?

2. If ∠B measures 12 degrees, what is the measure of its supplement?

3. If ∠C measures 50 degrees and it is bisected, what is the measure of the two resulting angles?

Look at these angles.

4. If ∠1 measures 30 degrees, what is the measure of ∠2?

5. If ∠1 measures 30 degrees, what is the measure of ∠3?

6. If a straight angle is bisected, what is the measure of each of the resulting angles?

(Answers are on page 34.)

Angle Congruence

Two figures with exactly the same size and shape are congruent. How do you determine if two angles are congruent? Two angles are congruent if any one of the following conditions are met:

- They have the same measure.
- You can put one on top of the other, and they are identical.
- They are both right angles since both measure 90 degrees.

- They are complements of the same angle. Complements of the same angle always have the same measure.
- They are supplements of the same angle. Supplements of the same angle always have the same measure.
- They are both congruent to the same angle. If two angles are congruent to the same angle, they have the same measure as that angle and they must be congruent to each other.

To show that two angles are congruent, write $\angle A \cong \angle B$. The symbol for congruence is a wavy line over an equals sign.

CAUTION—Major Mistake Territory!

Make sure that you know the difference between *equal* and *congruent*.

If $m\angle 1 = 32$ and $m\angle 2 = 32$, then the $m\angle 1$ is equal to the $m\angle 2$, but $\angle 1$ is not equal to $\angle 2$.

Two angles cannot be equal, but two angles can be congruent. Angle 1 is congruent to angle 2.

BRAIN TICKLERS   Set # 9

Look at the following diagram. Which angles are congruent to each other?

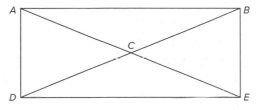

(Answers are on page 35.)

Word Problems

It's possible to use your skills in algebra and what you are learning about geometry to solve word problems.

Example:

If an angle is twice as large as its complement, what is the measure of the angle and its complement?

Change this problem to an equation.

The phrases in parentheses are changed to mathematical language.

x (an angle)
$=$ (is)
2 (twice)
$90 - x$ (its complement)

Now write an equation.

$$x = 2(90 - x)$$

Solve this equation.

$$x = 2(90 - x)$$
$$x = 180 - 2x$$

Add $2x$ to both sides.

$$3x = 180$$
$$x = 60$$

The angle is 60 degrees; its complement is 30 degrees.

Example:

An angle is 50 degrees less than its supplement. What is the measure of the angle and its supplement?

Change the problem to an equation. Watch as the phrases in parentheses are changed to mathematical language.

x (an angle)
$=$ (is)
$180 - x$ (its supplement)
$- 50$ (50 degrees less than)

Now put all these expressions together to form an equation.

$$x = 180 - x - 50$$

Now solve this equation.

First simplify both sides of the equation.

$$x = 180 - x - 50$$

Simplified, this equation becomes $x = 130 - x$.
Simplify both sides of the equation further to get $2x = 130$.
Divide both sides by 2, and the result is $x = 65$.
The angle is 65 degrees and its supplement is 115 degrees.

SUPER BRAIN TICKLERS

How well do you know your geometry terms? Fill in the blanks with angle language.

1. _____ angles measure exactly 90 degrees.

2. The sum of two _____ angles is exactly 180 degrees.

3. Two angles with exactly the same measure are _____.

4. An angle with a measure of less than 90 degrees is a(n) _____ angle.

5. The sum of two _____ angles is exactly 90 degrees.

6. _____ angles are formed when two lines intersect.

7. _____ angles measure greater than 90 degrees and less than 180 degrees.

8. A ray that divides an angle into two congruent angles is called an angle _____.

9. An angle that measures exactly 180 degrees is a(n) _____ angle.

(Answers are on page 35.)

BRAIN TICKLERS—THE ANSWERS

Set # 4, page 18

1. True

2. True

3. True

4. False

Set # 5, page 21

1. 15 degrees

2. 150 degrees

3. 90 degrees

4. 110 degrees

Set # 6, page 23

1. 95 degrees

2. 5 degrees

3. 50 degrees

Set # 7, page 25

1. True

2. False

3. True

4. False

Set # 8, page 30

1. 31 degrees

2. 168 degrees

3. 25 degrees, 25 degrees

4. 150 degrees

5. 30 degrees

6. 90 degrees, 90 degrees

Set # 9, page 31

Angle $ACD \cong$ angle BCE

Angle $ACB \cong$ angle DCE

Angle $BED \cong$ angle $EDA \cong$ angle $DAB \cong$ angle ABE

Super Brain Ticklers, page 33

1. Right

2. supplementary

3. congruent

4. acute

5. complementary

6. Vertical

7. Obtuse

8. bisector

9. straight

Chapter 3

Parallel and Perpendicular Lines

Points, lines, and planes are the primary elements of geometry. This chapter will explore the relationship between the lines in a plane. If two lines are in a plane, either they intersect or they are parallel.

EXPERIMENT

Discover the properties of lines in a plane.

Materials
  Two pencils
  A table

Procedure
  1. Place two pencils on a table. The table represents a plane.

  2. Let the two pencils represent two straight lines, but remember that straight lines extend infinitely. Make an X with the pencils. The pencils now intersect at exactly one point.

  3. Try to make the pencils intersect at more than one point. It's impossible! The only way to make two pencils intersect at more than one point is to put them on top of each other, but then they represent the same line.

Something to think about . . .
How many different ways can two lines intersect?

POSTULATE If two lines intersect, they intersect at exactly one point.

Example:

These lines intersect at exactly one point, *A*.

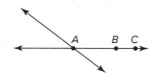

However, two lines can intersect an infinite number of different ways. These lines could also intersect at point *B*, point *C*, or any other point along either line.

If two lines intersect they form four angles.

When two lines intersect, the sum of any two adjacent angles is 180 degrees.

Example:

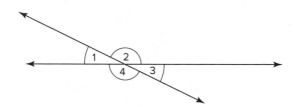

Angle 1 and angle 2 are adjacent angles and their sum is 180 degrees.
Angle 1 and angle 4 are adjacent angles and their sum is 180 degrees.
Angle 2 and angle 3 are adjacent angles and their sum is 180 degrees.
Angle 3 and angle 4 are adjacent angles and their sum is 180 degrees.

The sum of all four angles formed by two intersecting lines is 360 degrees.

Example:

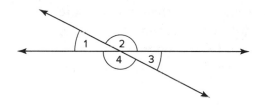

The sum of angles 1, 2, 3, and 4 is 360 degrees.

Vertical Angles

If two lines intersect, they form four angles. The opposite angles formed are called *vertical angles*.

Example:

Angles *a* and *c* are vertical angles. Angles *b* and *d* are vertical angles.

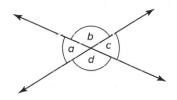

THEOREM   Vertical angles are always congruent. Vertical angles always have the same measure.

Example:

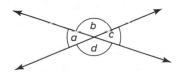

Angles *a* and *c* are vertical angles; therefore . . .
  angles *a* and *c* are congruent.
  angles *a* and *c* have the same measure.
Angles *b* and *d* are vertical angles; therefore . . .
  angles *b* and *d* are congruent.
  angles *b* and *d* have the same measure.

BRAIN TICKLERS   Set # 10

Use this diagram to solve the problems that follow.

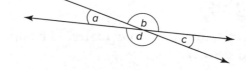

1. If m∠*a* = 15, what is the measure of angle *b*?

2. If m∠*a* = 15, what is the measure of angle c?

3. If m∠*a* = 15, what is the m∠*d*?

4. What is the sum of the m∠*a* and the m∠*b*?

5. What is the sum of the m∠*b* and the m∠*c*?

6. What is the sum of the m∠*c* and the m∠*d*?

7. What is the sum of the m∠*d* and the m∠*a*?

8. What is the sum of the measures of angles *a*, *b*, *c*, and *d*?

(Answers are on page 55.)

Perpendicular Lines

Sometimes two lines intersect to form right angles.

THEOREM   Perpendicular lines always intersect to form four right angles.

Example:

These lines are perpendicular to each other.

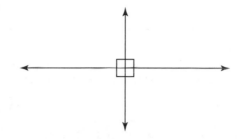

All four of these angles are right angles. The sum of all four of these angles is $90 + 90 + 90 + 90 = 360$. The opposite angles are vertical angles and are congruent. All four angles are congruent to each other.

EXPERIMENT

Discover the number of lines that can be drawn from a point not on the line to a line.

Materials
Red pencil
Black pencil
Ruler
Paper

Procedure
1. Draw a red line on a piece of paper.

2. Use the black pencil to place a black point above the red line.

3. How many lines can you draw through the black point that also intersect the red line?

Something to think about . . .
How many lines can be drawn perpendicular to the red line that go through a single point?

POSTULATE   Given a line and a point not on that line, there is exactly one line that passes through the point perpendicular to the line.

EXPERIMENT

Explore the relationship between a line and a point on the line.

Materials
  Red pencil
  Black pencil
  Paper
  Ruler

Procedure
  1. Draw a red line on a piece of paper. Make a black dot on the red line.

  ⟵———————•———————⟶

  2. How many lines can you draw through the black dot?

  3. Draw a second red line on a piece of paper. Make a black dot on this line.

  4. How many perpendicular lines can you draw through the black dot?

Something to think about . . .
How many perpendicular lines can you draw through two parallel lines?

POSTULATE   Through a point on a line, there is exactly one line perpendicular to the given line.

THEOREM   If two lines intersect and the adjacent angles are congruent, then the lines are perpendicular.

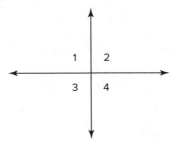

MINI-PROOF

How can you prove the theorem: *If two lines intersect and the adjacent angles are congruent, then the lines are perpendicular?*

What do you know?

1. Angle 1 and angle 2 are adjacent angles.

2. Angle 1 and angle 2 are congruent.

3. The measure of angle 1 plus the measure of angle 2 is equal to a straight angle, which is 180.

What does this mean?

4. Since angles 1 and 2 are congruent, the measures of angles 1 and 2 are equal.

5. Since angle 1 is equal to angle 2 and they form a straight angle, they must each be equal to 90 degrees.

6. Since angle 1 and angle 2 are each 90 degrees, they are each right angles.

What can you conclude?

7. Since both angle 1 and angle 2 are right angles, the lines must be perpendicular.

BRAIN TICKLERS   Set # 11

Determine whether each of these statements is true or false.

1. Two intersecting lines are always perpendicular.

2. Two different lines can intersect at more than one point.

3. Perpendicular lines form acute angles.

4. The sum of the measures of the angles of two perpendicular lines is 360 degrees.

5. All right angles are congruent to each other.

(Answers are on page 55.)

Parallel Lines

Two lines are parallel if and only if they are in the same plane and they never intersect. Parallel lines are the same distance from each other over their entire lengths.

If two lines are cut by a transversal, eight angles are formed. A *transversal* is a line that intersects two lines at different points.

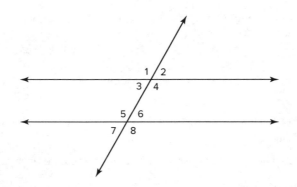

These angles have specific names. Memorize them!

- *Exterior angles* are angles that lie outside the space between the two lines. Angles 1, 2, 7, and 8 are *exterior angles*.

- *Interior angles* lie in the space between the two lines. Angles 3, 4, 5, and 6 are *interior angles*.

- *Alternate interior angles* are angles that lie between the two lines but on opposite sides of the transversal. Angles 4 and 5 are *alternate interior angles*. Angles 3 and 6 are *alternate interior angles*.

- *Alternate exterior angles* are angles that lie outside the two lines and on opposite sides of the transversal. Angles 1 and 8 are *alternate exterior angles*. Angles 2 and 7 are *alternate exterior angles*.

- *Corresponding angles* are nonadjacent angles on the same side of the transversal. One corresponding angle must be an *interior angle* and one must be an *exterior angle*. Angles 1 and 5 are *corresponding angles*. Angles 2 and 6 are *corresponding angles*. Angles 3 and 7 are *corresponding angles*. Angles 4 and 8 are *corresponding angles*.

- *Consecutive interior angles* are nonadjacent interior angles that lie on the same side of the transversal. Angles 3 and 5 are *consecutive interior angles*. Angles 4 and 6 are *consecutive interior angles*.

BRAIN TICKLERS   Set # 12

Determine the relationship between each of the following pairs of angles. Circle the correct letter(s) for each pair.

A = Adjacent
AI = Alternate Interior
AE = Alternate Exterior
C = Corresponding Angles
V – Vertical Angles
S = Supplementary Angles

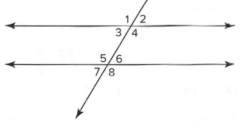

1. ∠1 and ∠2: A AI AE C V S
2. ∠1 and ∠3: A AI AE C V S
3. ∠1 and ∠4: A AI AE C V S
4. ∠1 and ∠5: A AI AE C V S
5. ∠1 and ∠6: A AI AE C V S
6. ∠1 and ∠7: A AI AE C V S
7. ∠1 and ∠8: A AI AE C V S

(Answers are on page 55.)

PARALLEL POSTULATE   Given a line and a point not on the line, there is only one line parallel to the given line.

Draw a line. Draw a dot, not on the line. Label it P.

• P

How many lines can you draw through the dot parallel to the line?

POSTULATE   If two lines are cut by a transversal and the corresponding angles are equal, then the lines are parallel.

THEOREM   If two parallel lines are cut by a transversal, then their alternate interior angles are congruent.

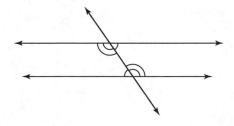

What does the term *alternate interior* mean?

Alternate means on different sides of the transversal.
Interior means between the two parallel lines.
Alternate interior means between the two parallel lines and on opposite sides of the transversal.

Example:

Look at these two parallel lines.

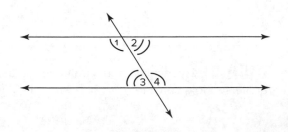

Angles 1 and 4 are alternate interior angles.
Angles 2 and 3 are alternate interior angles.
The measure of angle 1 is equal to the measure of angle 4.
The measure of angle 2 is equal to the measure of angle 3.

Remember, when two parallel lines are cut by a transversal, eight different angles are formed. If you know the measure of one of the eight angles, you can find the measure of all eight of the angles.

Example:

If the measure of angle 3 is 60 degrees, you can find the measure of all the rest of the angles.

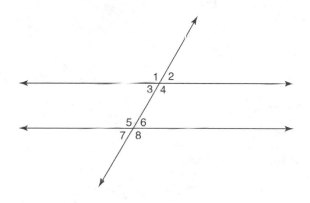

The measure of angle 2 is also 60 degrees, since it is vertical to angle 3.
The measure of angle 6 is also 60 degrees, since it is an alternate interior angle to angle 3.
The measure of angle 7 is also 60 degrees, since it is a vertical angle to angle 6.
The measure of angle 1 is 120 degrees, since it is supplemental to angle 3.
The measure of angle 4 is 120 degrees, since it is vertical to angle 1.
The measure of angle 5 is 120 degrees, since it is an alternate interior angle to angle 4.
The measure of angle 8 is 120 degrees, since it is a vertical angle to angle 5.

BRAIN TICKLERS   Set # 13

Assume that the measure of angle 4 is 110 degrees.

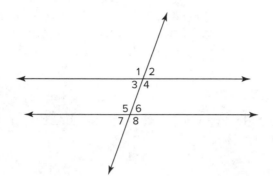

1. What is the m∠1?

2. What is the m∠2?

3. What is the m∠3?

4. What is the m∠5?

5. What is the m∠6?

6. What is the m∠7?

7. What is the m∠8?

(Answers are on page 56.)

More on Parallel Lines

THEOREM   If two parallel lines are cut by a transversal, then their corresponding angles are congruent.

What are *corresponding angles*? Corresponding angles are angles that lie on the same side of the transversal. One corresponding angle lies on the interior of the parallel lines while the other corresponding angle lies on the exterior of the parallel lines.

Example:

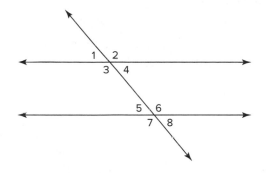

Angles 1 and 5 are corresponding angles, so they are congruent.
Angles 2 and 6 are corresponding angles, so they are congruent.
Angles 3 and 7 are corresponding angles, so they are congruent.
Angles 4 and 8 are corresponding angles, so they are congruent.

THEOREM   If two parallel lines are cut by a transversal, their alternate exterior angles are congruent.

Exterior angles are found above or below the pair of parallel lines.
Alternate angles lie on opposite sides of the transversal.
Alternate exterior angles lie outside the parallel lines, and on the opposite sides of the transversal.

Example:

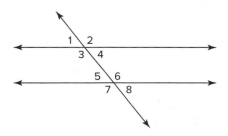

Angles 1 and 8 are alternate exterior angles; therefore, they are congruent.
Angles 2 and 7 are alternate exterior angles; therefore, they are congruent.

THEOREM   If two parallel lines are cut by a transversal, the consecutive interior angles are supplementary.

The *consecutive interior angles* are angles on the same side of the transversal and inside both parallel lines.

Example:

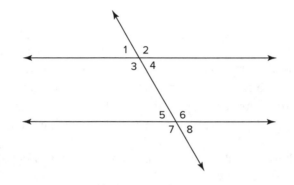

Angles 3 and 5 are consecutive interior angles.
Angles 4 and 6 are consecutive interior angles.
Angles 3 and 5 are supplementary angles.
Angles 4 and 6 are supplementary angles.

EXPERIMENT

Explore the relationship between angles formed by parallel lines and a transversal.

Materials
Pencil
Paper
Ruler
Protractor

Procedure
1. Draw two parallel lines.

2. Draw a transversal across the lines.

3. Label the eight angles formed angles 1, 2, 3, 4, 5, 6, 7, and 8.

4. Measure each of the angles and write its measure on the diagram.

5. Describe the relationship between each pair of angles as complementary, supplementary, equal, and unknown. Enter the results in the chart. Place a *C* for Complementary, *S* for Supplementary, *E* for Equal, and *U* for Unknown.

	Angle 1	Angle 2	Angle 3	Angle 4	Angle 5	Angle 6	Angle 7	Angle 8
Angle 1								
Angle 2								
Angle 3								
Angle 4								
Angle 5								
Angle 6								
Angle 7								
Angle 8								

Something to think about . . .

What did you notice about the relationship of the angles?

Sum it up!

When two parallel lines are cut by a transversal, the following pairs of angles are congruent:

- alternate interior angles
- corresponding angles
- alternate exterior angles

When two parallel lines are cut by a transversal, the following pairs of angles are supplementary:

- consecutive interior angles

THEOREM   If two parallel lines are cut by a transversal, then any two angles are either congruent or supplementary.

THEOREM   If a line is perpendicular to one of two parallel lines, then it is perpendicular to the other.

Proving lines parallel

There are four ways to prove two lines parallel. First, cut the two lines by a transversal. If any of the following are true, then the lines are parallel.

1. The alternate interior angles are congruent.

2. Their corresponding angles are congruent.

3. Their alternate exterior angles are congruent.

4. The interior angles on the same side of the transversal are supplementary.

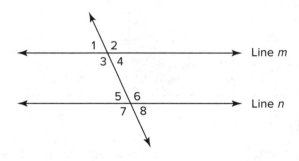

Way 1—Alternate Interior Angles

If ∠4 and ∠5 are congruent, then line *m* is parallel to line *n*.
If ∠3 and ∠6 are congruent, then line *m* is parallel to line *n*.

Way 2—Corresponding Angles

If ∠1 and ∠5 are congruent, then line *m* is parallel to line *n*.
If ∠2 and ∠6 are congruent, then line *m* is parallel to line *n*.
If ∠3 and ∠7 are congruent, then line *m* is parallel to line *n*.
If ∠4 and ∠8 are congruent, then line *m* is parallel to line *n*.

Way 3—Alternate Exterior Angles

If ∠1 and ∠8 are congruent, then line *m* is parallel to line *n*.
If ∠2 and ∠7 are congruent, then line *m* is parallel to line *n*.

BRAIN TICKLERS   Set # 14

Are these two lines parallel?

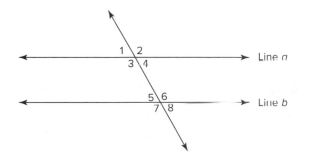

If each of the following equations are true, decide if line *a* and line *b* are parallel. Answer *Yes, No,* or *Don't know*.

1. m∠1 = m∠5

4. m∠4 = m∠5

2. m∠1 = m∠4

5. m∠1 = m∠8

3. m∠1 = m∠3

(Answers are on page 56.)

PAINLESS TIP

Draw a pair of parallel lines and a transversal on nine index cards. Using small arcs, draw one pair of angles on each card. Make sure to draw one card with a pair of vertical angles, adjacent angles, supplementary angles, complementary angles, interior angles, alternate interior angles, exterior angles, alternate exterior angles, and corresponding angles. Label the name of the angle on the back of each card. Use the cards to memorize the types of angles.

 **SUPER BRAIN TICKLERS**

Fill in the blanks with the correct terms to find out how well you understand perpendicular and parallel lines.

1. If two lines intersect, they form _____ angles.

2. Vertical angles are always _____.

3. Parallel lines never _____.

4. Perpendicular lines intersect to form four _____ angles.

5. If two parallel lines are cut by a transversal, then three types of angles are equal. What are they?

    a. _____

    b. _____

    c. _____

6. If two parallel lines are cut by a transversal, then any two angles are either congruent or _____.

7. Through a point not on a line there is (are) _____ perpendicular line(s) to the given line.

(Answers are on page 56.)

BRAIN TICKLERS—THE ANSWERS

Set # 10, page 40

1. 165
2. 15
3. 165
4. 180
5. 180
6. 180
7. 180
8. 360

Set # 11, page 43

1. False
2. False
3. False
4. True
5. True

Set # 12, page 45

1. A, S
2. A, S
3. V
4. C
5. S
6. S
7. AE

Set # 13, page 48

1. 110

2. 70

3. 70

4. 110

5. 70

6. 70

7. 110

Set # 14, page 53

1. Yes

2. Don't know

3. Don't know

4. Yes

5. Yes

Super Brain Ticklers, page 54

1. four

2. congruent

3. intersect

4. right

5. a. Corresponding

   b. Alternate exterior

   c. Alternate interior

6. supplementary

7. one

Chapter 4

Triangles

Interior Angles

A *polygon* is a closed figure with three or more sides. The sides of a polygon are line segments. Each side of a polygon intersects two other sides of the polygon at their endpoints. The endpoints of the sides of a polygon are the vertices of the polygon. A polygon lies in a single plane.

A *triangle* is the simplest type of polygon. A triangle has three sides and three angles. Each side of a triangle is labeled by its endpoints.

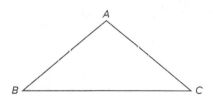

The sides of this triangle are $\overline{AB}$, $\overline{BC}$, and $\overline{CA}$.
When labeling the sides of a triangle, it doesn't matter which letter comes first.

Side $\overline{AB}$ could also be labeled $\overline{BA}$.
Side $\overline{BC}$ could also be labeled $\overline{CB}$.
Side $\overline{CA}$ could also be labeled $\overline{AC}$.

Each two sides of a triangle form an angle. Where any two sides of a triangle meet is a *vertex* of the triangle. A triangle has three vertices.

A triangle has three *angles*, which are actually *interior angles*. Each of the interior angles of a triangle is less than 180 degrees. An angle of 180 degrees is a straight line and cannot form a triangle.

The angles of a triangle are labeled three different ways.

1. The interior angles of a triangle can be labeled using three letters. The center letter is the vertex of the angle. The marked angle is ∠*ABC*. It could also be labeled ∠*CBA*.

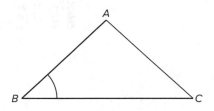

In this triangle, the marked angle is ∠*BAC* or ∠*CAB*.

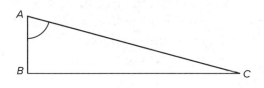

2. The interior angle can be labeled by the single point that is the vertex of the angle, if and only if no other angles share the same vertex. The angles of this triangle can be labeled ∠*A*, ∠*B*, and ∠*C*.

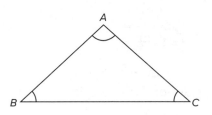

3. The interior angles can also be labeled by placing a small letter or number in the vertex of the angle. The angles of this triangle can be labeled ∠*a*, ∠*b*, and ∠*c*.

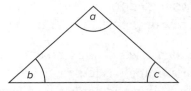

There are three ways to label the same angle. Look at triangle *XYZ* below.

∠*YXZ*, ∠*X*, and ∠1 are all the same angle.
∠*XYZ*, ∠*Y*, and ∠2 are all the same angle.
∠*XZY*, ∠*Z*, and ∠3 are all the same angle.

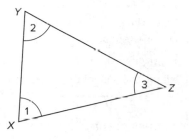

EXPERIMENT

Find the sum of the angles of a triangle.

Materials
Paper
Pencil
Ruler
Scissors

Procedure
1. Draw a triangle on a piece of paper.

2. Cut out the triangle.

3. Rip off the three angles of the triangle.

4. Put the three angles in a row so that the angles meet at one point and at least one side of each angle touches the side of another angle.

They form a straight line. The sum of the angles of a triangle is 180 degrees.
5. Draw a second triangle and repeat Steps 2 and 3.

Something to think about . . .
Can you draw a triangle where the sum of the angles is not 180 degrees?

THEOREM   The sum of the measures of the interior angles of a triangle is 180 degrees.

EXPERIMENT

Measure the angles of a triangle.

Materials
Protractor
Pencil
Paper

Procedure
1. Draw five triangles on a piece of paper. Make the triangles as different as you can.
2. Label the triangles One, Two, Three, Four, Five.
3. Using a protractor, measure the angles of each triangle. Enter the results in the chart.
4. Add the measurements of each of the angles of each triangle together. Put the results in the chart.

Triangles	Angle 1	Angle 2	Angle 3	Sum of Angles
One				
Two				
Three				
Four				
Five				

Did the sum of the angles of each of your triangles add up to 180 degrees?

Something to think about . . .
If you add the measures of the angles of a square, what is the sum? Does every square have the same number of degrees?

__THEOREM__   Every angle of a triangle has a measure greater than 0 and less than 180 degrees.

If you add the measures of the angles of any triangle together, the answer will always be 180 degrees.

If you know the measure of any two angles of a triangle, follow these two painless steps to compute the measure of the third angle.

Step 1: Find the sum of the two angles you know.

Step 2: Subtract this sum from 180 degrees.

The result is the measure of the third angle.

Example:

If a triangle has angles measuring 60 degrees and 40 degrees, what is the measure of the third angle?

Step 1: Add the two angles.

$$60 + 40 = 100$$

Step 2: Subtract the sum from 180.

$$180 - 100 = 80$$

The third angle is 80 degrees.

Example:

Find the measure of a third angle of a right triangle that has an acute angle measuring 40 degrees.

Step 1: Add the two angles.

Since it is a right triangle, one angle is 90 degrees.

$$90 + 40 = 130$$

Step 2: Subtract the sum from 180.

$$180 - 130 = 50$$

The third angle is 50 degrees.

BRAIN TICKLERS   Set # 15

Find the measure of the missing angle in each of these triangles.

1. What is the measure of angle *C* in triangle *ABC* if the measure of angle *A* is 70 degrees and angle *B* is 80 degrees?

2. If the measure of angle *A* is 150 degrees and angle *B* is 20 degrees, what is the measure of angle *C* in triangle *ABC*?

3. If the measure of one acute angle of a right triangle is 10 degrees, what is the measure of the other acute angle?

(Answers are on page 92.)

Exterior Angles

Triangles can have both interior and exterior angles. Each exterior angle of a triangle forms a straight line with an interior angle of the triangle. If you extend one side of a triangle, the angle formed is an exterior angle.

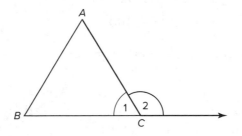

The interior angle is angle 1. Angle 2 is the exterior angle. Angle 1 is inside the triangle, while angle 2 is outside the triangle. Notice that the interior angle and the exterior angle form a straight line. The sum of the m∠1 and the m∠2 equals 180 degrees. Angle 1 and angle 2 are supplementary angles.

At each interior angle of a triangle, you can actually form *two* exterior angles. Extend both sides of the triangle at point *C* to form two exterior angles.

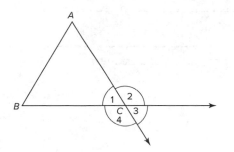

Angle 1 is the original interior angle.
Angle 2 is an exterior angle.
Angle 3 is a vertical angle formed by the two intersecting lines.
Angle 4 is the other exterior angle.

Notice the relationships between these four angles.

Angle 1 and angle 2 are supplementary. The sum of their
measures equals 180 degrees.
Angle 1 and angle 4 are also supplementary. The sum of their
measures equals 180 degrees.
Angle 1 and angle 3 are vertical angles. They are congruent.
Angle 2 and angle 4 are vertical angles. They are congruent.

Every triangle has six possible exterior angles. In this diagram, all the
exterior angles of one triangle are drawn. Name the exterior angles.
How many pairs of vertical angles can a triangle have?

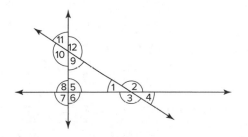

The exterior angles of this triangle are 2, 3, 6, 8, 10, and 12.
This triangle has six pairs of vertical angles.

EXPERIMENT

Explore the exterior angles of a triangle.

Materials
Protractor
Pencil
Paper
Ruler

Procedure

1. Draw an acute triangle. Extend one of the sides of the triangle to form an exterior angle.

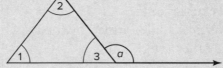

2. Label the exterior angle *a*. Label the interior angles 1, 2, and 3. In the above triangle,

    Angle *a* is the exterior angle.
    Angle 3 is the adjacent interior angle.
    Angles 1 and 2 are the nonadjacent interior angles.

3. Measure the exterior angle. Enter your answer in the chart.

4. Measure each of the nonadjacent interior angles. Enter your answers in the chart.

5. Add the measures of angles 1 and 2. Enter the result in the chart.

6. Draw a right triangle.

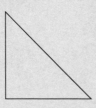

7. Extend one of the sides of the right triangle.

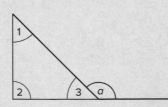

8. Label the exterior angle *a*. Label the interior angles 1, 2, and 3. In the above triangle,

   Angle *a* is the exterior angle.
   Angle 3 is the adjacent interior angle.
   Angles 1 and 2 are the nonadjacent interior angles.

9. Measure the exterior angle. Enter the answer in the chart.

10. Measure each of the nonadjacent interior angles. Enter the results in the chart.

11. Add the measures of angles 1 and 2. Enter the result in the chart.

12. Draw an obtuse triangle.

13. Extend one of the sides of the obtuse triangle.

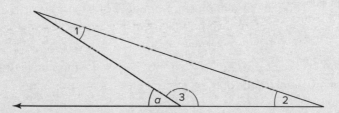

14. Label the exterior angle *a*. Label the interior angles 1, 2, and 3. In the above triangle,

    Angle *a* is the exterior angle.
    Angle 3 is the adjacent interior angle.
    Angles 1 and 2 are the nonadjacent interior angles.

15. Measure the exterior angle. Enter your answer in the chart.

16. Measure each of the nonadjacent interior angles. Enter the results in the chart.

17. Add the measures of angles 1 and 2. Enter your answer in the chart.

Type of Triangle	Measure of exterior angle *a*	Measure of adjacent interior angle 3	Measure of non-adjacent interior angle 1	Measure of non-adjacent interior angle 2	Sum of angles 1 and 2
Acute triangle					
Right triangle					
Obtuse triangle					

Do any two columns of the chart always match? Which ones?

Something to think about . . .

What is the sum of all six exterior angles of a triangle? Is it the same for every triangle?

THEOREM   The measure of an exterior angle of a triangle is equal to the sum of the measures of the two nonadjacent interior angles.

THEOREM   The measure of any exterior angle of a triangle is greater than the measure of either of the two nonadjacent interior angles.

BRAIN TICKLERS   Set # 16

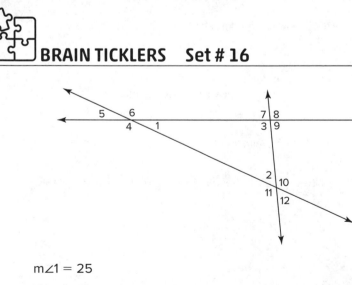

m∠1 = 25

m∠2 = 60

What is the measure of these angles?

1. ∠3

2. ∠4

3. ∠5

4. ∠6

5. ∠7

6. ∠8

7. ∠9

8. ∠10

9. ∠11

10. ∠12

(Answers are on page 92.)

Types of Triangles

Triangles are labeled by the size of their angles. There are acute, obtuse, and right triangles.

Acute triangles

An *acute triangle* is a triangle in which every angle has measures less than 90 degrees. Every angle of an acute triangle is an acute angle.

Here are three acute triangles.

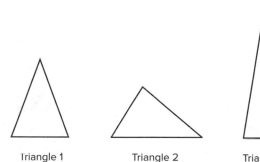

Triangle 1   Triangle 2   Triangle 3

BRAIN TICKLERS   Set # 17

Label each statement true or false.

1. A triangle with angles of 50, 60, and 70 degrees is an acute triangle.

2. The sum of the angles of an acute triangle is less than 180 degrees.

3. A triangle with angles of 10 and 20 degrees is an acute triangle.

(Answers are on page 92.)

Obtuse triangles

An *obtuse triangle* has one obtuse angle. An obtuse angle is an angle that is greater than 90 degrees.

BRAIN TICKLERS   Set # 18

Label each statement true or false.

1. A triangle with angles of 10, 20, and 150 degrees is an obtuse triangle.

2. A triangle with two 40-degree angles is an obtuse triangle.

3. An obtuse triangle can have two obtuse angles.

(Answers are on page 92.)

Right triangles

A *right triangle* has one right angle. A right triangle has one angle that measures 90 degrees.

In this right triangle, angle *ABC* is a right angle. The other two angles are acute angles. A triangle can have at most one right angle.

A right triangle has one right angle. The sides of the triangle that form the right angle are the legs of the triangle. The side opposite the right angle is called the hypotenuse of the triangle.

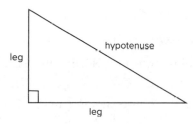

EXPERIMENT

Explore right triangles.

Materials

Protractor   Ruler
Pencil   Paper

Procedure

1. Draw five different right triangles. The legs of the triangles should be the length shown in the chart.

2. Measure the angles of each right triangle using a protractor.

3. Look for patterns in the chart.

Triangle	Leg 1	Leg 2	Measure of angle *A*	Measure of angle *B*	Measure of angle *C*	Measure of hypotenuse
Triangle 1	1 inch	2 inches	90			
Triangle 2	3 inches	6 inches	90			
Triangle 3	5 inches	5 inches	90			
Triangle 4	1 inch	7 inches	90			
Triangle 5	2 inches	2 inches	90			

Something to think about . . .

What is always true about the other two angles of a right triangle? (Hint: Find the sum of angle *B* and angle *C*.)

What other relationships can you discover about right triangles?

THEOREM   The hypotenuse of a right triangle is longer than either of the other two sides.

BRAIN TICKLERS   Set # 19

Answer each of these statements true or false.

1. A right triangle can have two 60-degree angles.

2. A triangle with angles measuring 30, 60, and 90 degrees is a right triangle.

3. A triangle with two 45-degree angles is a right triangle.

(Answers are on page 93.)

More Types of Triangles

Another way to classify triangles is to look at the number of congruent angles in a triangle. Some triangles have two congruent angles, some have three congruent angles, and in some triangles, none of the angles are congruent.

Isosceles triangles

An *isosceles triangle* has two congruent sides.

In this triangle, sides *AB* and *AC* are congruent.

THE BASE ANGLES   If two sides of a triangle are congruent, the angles opposite these sides are congruent.

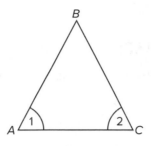

In this triangle, if $\overline{AB} \cong \overline{BC}$, then $\angle 1 \cong \angle 2$.

EXPERIMENT

Measure the angles of an isosceles triangle.

Materials
Paper   Protractor
Pencil   Scissors

Procedure

1. Fold a piece of paper in half.

2. Cut a triangle out of the piece of paper along the fold by cutting a "v" in the paper with one cut perpendicular to the fold. Open the folded piece of paper. It's an isosceles triangle.

3. Cut out three more isosceles triangles. Make them all different sizes and shapes. Measure the congruent angles of each isosceles triangle.

Triangle	Measure of congruent angles
Triangle 1	
Triangle 2	
Triangle 3	
Triangle 4	

Something to think about . . .

How many of the congruent angles are greater than 90 degrees?

How many of the congruent angles are less than 90 degrees?

How many of the congruent angles are equal to 90 degrees?

BRAIN TICKLERS   Set # 20

Decide whether each of these statements is true or false.

1. A triangle with 30-, 60-, and 90-degree angles is an isosceles triangle.

2. An isosceles triangle can have two 90-degree angles.

3. A triangle with angles that measure 40 degrees and 100 degrees is an isosceles triangle.

(Answers are on page 93.)

Equiangular triangles

An *equiangular triangle* has three congruent angles.

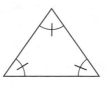

Each angle of an equiangular triangle is 60 degrees. There are always 180 degrees in a triangle. If you divide 180 degrees by 3, the result is 60.

Equilateral triangles

A triangle with three equal sides is an *equilateral triangle*. An equilateral triangle is equiangular.

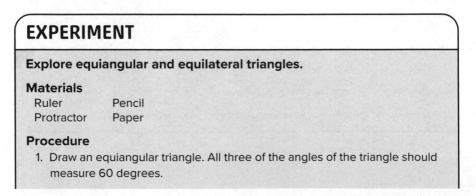

EXPERIMENT

Explore equiangular and equilateral triangles.

Materials
Ruler   Pencil
Protractor   Paper

Procedure
1. Draw an equiangular triangle. All three of the angles of the triangle should measure 60 degrees.

2. Measure the sides of the triangle. Are all three sides of the triangles equal? If a triangle has three equal sides, it is an equilateral triangle.

3. Draw a second equiangular triangle. Make sure all three angles of this triangle equal 60 degrees.

4. Measure the sides of this second triangle. Are all three sides of this second triangle equal?

5. Draw a triangle with three 4-inch sides.

6. Measure the angles of this triangle using a protractor. What is the measure of each of the angles of this triangle?

Something to think about . . .

Is every equiangular triangle an equilateral triangle?

Is every equilateral triangle an equiangular triangle?

 **BRAIN TICKLERS   Set # 21**

Determine whether each of the following statements is true or false.

1. A triangle with three 60-degree angles is an equiangular triangle.

2. A triangle with two 60-degree angles is an equiangular triangle.

3. A triangle with three 40-degree angles is an equiangular triangle.

(Answers are on page 93.)

Scalene triangles

A *scalene triangle* has no congruent sides and no congruent angles.

A scalene triangle can be an acute, an obtuse, or a right triangle. A triangle whose angles measure 100 degrees, 45 degrees, and 35 degrees is an obtuse triangle, because it contains an obtuse angle. It is also a scalene triangle, since the measurement of all three angles is different.

Similarly, a triangle with angles of 90, 60, and 30 degrees is a right triangle and a scalene triangle.

EXPERIMENT

Explore angles and sides of scalene triangles.

Materials
Paper
Pencil
Ruler

Procedure

1. Draw a scalene triangle. Label the angles of $\triangle ABC$.

2. Measure the angles of the triangles.

3. Rank order the angles of the triangle from smallest to largest. Enter the results in the chart.

4. Measure the sides of the triangle.

5. Rank order the sides of the triangle from smallest to largest. Enter the results in the chart.

Name of angle	Rank of angle measure	Name of side	Rank of side length
Angle A		AB	
Angle B		BC	
Angle C		AC	

6. Draw another scalene triangle. Repeat Steps 1 to 5.

Name of angle	Rank of angle measure	Name of side	Rank of side length
Angle A		AB	
Angle B		BC	
Angle C		AC	

Something to think about . . .

Is there a relationship between the largest angle and the longest side?

Is there a relationship between the smallest angle and the shortest side?

THEOREM   The side of a triangle opposite the largest angle of a triangle is the largest side of the triangle.

BRAIN TICKLERS   Set # 22

Decide whether each of the following statements is true or false.

1. A triangle with angles measuring 50, 60, and 70 degrees is a scalene triangle.

2. A triangle with angles measuring 170 and 2 degrees is a scalene triangle.

3. A triangle with angles measuring 45 and 90 degrees is a scalene triangle.

4. A right triangle with a 30-degree angle is a scalene triangle.

(Answers are on page 93.)

BRAIN TICKLERS   Set # 23

Here are the measurements of two angles of various triangles. What type of triangles are these? Mark each set of measurements with the letters that apply.

A = Acute triangle
E = Equilateral triangle
I = Isosceles triangle
O = Obtuse triangle
R = Right triangle
S = Scalene triangle

1. Angles: 90 degrees, 45 degrees

2. Angles: 80 degrees, 40 degrees

3. Angles: 90 degrees, 60 degrees

4. Angles: 60 degrees, 60 degrees

5. Angles: 75 degrees, 75 degrees

6. Angles: 30 degrees, 45 degrees

7. Angles: 110 degrees, 35 degrees

8. Angles: 10 degrees, 70 degrees

(Answers are on page 93.)

Perimeter of a Triangle

The *perimeter* of a triangle is the distance around the triangle. To find the perimeter of a triangle, add the length of each of the three sides of the triangle together.

Here is the painless solution for finding the perimeter of a triangle.

To find the perimeter of a triangle, add the measurements of the sides.

Example:

What is the perimeter of a triangle with sides 4 centimeters, 5 centimeters, and 7 centimeters?

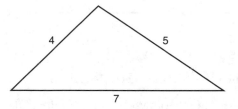

Add the three sides of the triangle:

$$4 \text{ cm} + 5 \text{ cm} + 7 \text{ cm} = 16 \text{ cm}$$

The perimeter is 16 centimeters.

Example:

Find the perimeter of an equilateral triangle if each of the sides of the triangle is 2 inches.

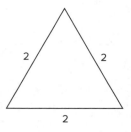

Add the three sides: 2 in. + 2 in. + 2 in. = 6 in.

BRAIN TICKLERS   Set # 24

1. What is the perimeter of a triangle with sides 4, 10, and 12?

2. What is the perimeter of an equilateral triangle with side 5?

3. What is the perimeter of an isosceles right triangle with hypotenuse length $\sqrt{8}$ and sides length 2?

(Answers are on page 94.)

Area of a Triangle

The *altitude* of a triangle is the line segment from the vertex of the triangle perpendicular to the opposite side.

Each triangle has three altitudes. The altitudes of this triangle are $\overline{AE}, \overline{CD}$, and $\overline{BF}$.

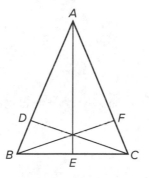

Two of the altitudes of this obtuse triangle are on the exterior of the triangle. The altitudes of this obtuse triangle are $\overline{AE}, \overline{BD}$, and $\overline{CF}$.

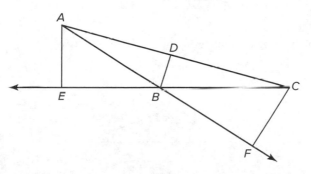

Two of the altitudes of a right triangle are the legs of the triangle. $\overline{AB}$ is one of the altitudes of this right triangle. $\overline{BC}$ is also an altitude of this right triangle. $\overline{BD}$ is the final altitude of this right triangle.

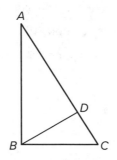

EXPERIMENT

See how the area of a triangle is related to the area of a rectangle.

Materials
Scissors
Graph paper
Pencil
Straight edge

Procedure
1. On a piece of graph paper, draw a right triangle 8 squares long and three squares tall.

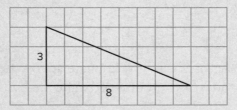

2. Draw another right triangle with the same height and width.

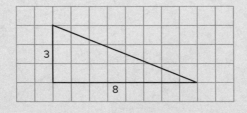

3. Cut out both triangles and fit them together.

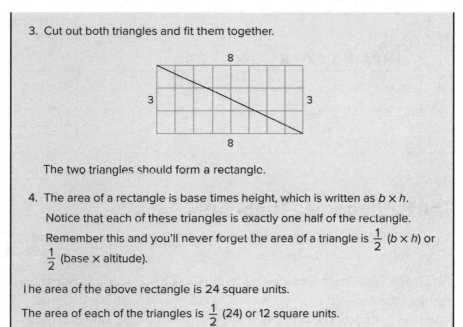

The two triangles should form a rectangle.

4. The area of a rectangle is base times height, which is written as $b \times h$.

Notice that each of these triangles is exactly one half of the rectangle. Remember this and you'll never forget the area of a triangle is $\frac{1}{2}$ ($b \times h$) or $\frac{1}{2}$ (base $\times$ altitude).

The area of the above rectangle is 24 square units.

The area of each of the triangles is $\frac{1}{2}$ (24) or 12 square units.

THEOREM   The area of a triangle is $\frac{1}{2}$ (base $\times$ altitude).

Example:

Find the area of a triangle with base 10 and altitude 1.

The area of a triangle is $\frac{1}{2}$ (base)(altitude).

The area is $\frac{1}{2}$ (10)(1).

The area is 5.

Example:

Find the area of a right triangle with legs 7 and 10.

Use the length of one of the legs of the triangle as the base of a triangle and use the length of the other leg as the altitude of the triangle.

The area is $\frac{1}{2}$ (10)(7), which is $\frac{1}{2}$ (70) = 35.

BRAIN TICKLERS   Set # 25

1. What is the area of a triangle with base 4 inches and altitude 6 inches?

2. What is the area of a right triangle with legs 5 and 8 inches long?

(Answers are on page 94.)

The Pythagorean Theorem

The square of the length of the hypotenuse of a right triangle is equal to the sum of the squares of the lengths of the legs.

Look at a right triangle. It has three sides. The side opposite the right angle is called the hypotenuse. The hypotenuse is a diagonal line that connects the other two sides of the triangle, called the legs.

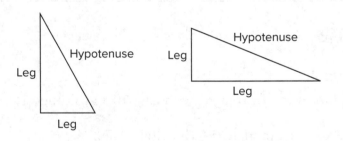

If you have a right triangle, you can determine the length of any side if you know the length of the other two sides.

If you know the length of both legs of a right triangle, just follow these three painless steps to find the length of the hypotenuse.

Step 1: Square the length of the legs of the triangle.

Step 2: Add the squares of the legs.

Step 3: Take the square root of the result of Step 2. The result is the length of the hypotenuse of the triangle.

Example:

If one leg of a right triangle is 3 inches long and the other leg of a right triangle is 4 inches long, how long is the hypotenuse?

Step 1: Square the length of legs. (Remember, to square a number, multiply the number by itself.)

Leg 1 is 3 inches. 3^2 is 9 square inches.
Leg 2 is 4 inches. 4^2 is 16 square inches.

Step 2: Add the squares of the legs.
$$9 + 16 = 25$$

Step 3: Take the square root of 25 square inches.

The square root of 25 square inches is 5 inches.
The length of the hypotenuse is 5 inches.

If you know the length of one leg and the hypotenuse of a right triangle, just follow these four painless steps to find the length of the other leg of the triangle.

Step 1: Square the length of the leg of the triangle.

Step 2: Square the length of the hypotenuse.

Step 3: Subtract the square of the leg of the triangle from the square of the hypotenuse.

Step 4: Find the square root of the result of Step 3.

Example:

Find the length of a leg of a right triangle if one leg is 12 inches and the hypotenuse is 13 inches.

Step 1: Find the square of the leg.
$$12^2 = 12 \times 12 = 144 \text{ square inches}$$

Step 2: Find the square of the hypotenuse.
$$13^2 = 13 \times 13 = 169 \text{ square inches}$$

Step 3: Subtract the square of the leg from the square of the hypotenuse.

$$169 - 144 = 25 \text{ square inches}$$

Step 4: Take the square root of the result of Step 3.

The square root of 25 is 5.
The length of the other leg is 5 inches.

The measurements of the sides of the triangle are 5, 12, and 13 inches.

Pythagorean Triples

A Pythagorean triple is a set of three positive integers that fit the Pythagorean equation.

THEOREM   The square of each of the two smaller integers is equal to the square of the largest integer: $a^2 + b^2 = c^2$.

To test if the set of three positive integers 3, 4, 5 is a Pythagorean triple, substitute 3, 4, and 5 into the Pythagorean equation $a^2 + b^2 = c^2$ and solve.

$$3^2 + 4^2 = 5^2$$
$$9 + 16 = 25$$
$$25 = 25$$

Since both sides of the equation are equal, 3, 4, 5 is a Pythagorean triple.

To test if the set of three positive integers 9, 40, 41 is a Pythagorean triple, substitute 9, 40, and 41 into the Pythagorean equation $a^2 + b^2 = c^2$ and solve.

$$9^2 + 40^2 = 41^2$$
$$81 + 1{,}600 = 1{,}681$$
$$1{,}681 = 1{,}681$$

Because both sides of the equation are equal, 9, 40, 41 is a Pythagorean triple.

To test if the set of three positive integers 4, 5, 6 is a Pythagorean triple, substitute 4, 5, 6 into the Pythagorean equation $a^2 + b^2 = c^2$ and solve.

$$4^2 + 5^2 = 6^2$$
$$16 + 25 = 36$$
$$41 \neq 36$$

Since 41 is not equal to 36, 4, 5, 6 is NOT a Pythagorean triple.

Multiples of Pythagorean triples

The multiple of a Pythagorean triple is a Pythagorean triple.

Example:

The set (3, 4, 5) is a Pythagorean triple. Multiply each of the numbers in the set by 2. The result (6, 8, 10) is a Pythagorean triple. Substitute the new set (6, 8, 10) into the Pythagorean equation
$a^2 + b^2 = c^2$ and solve to check.

$$6^2 + 8^2 = 10^2$$
$$36 + 64 = 100$$
$$100 = 100$$

Example:

Multiply the same Pythagorean triple (3, 4, 5) by 3. The result (9, 12, 15) is a Pythagorean triple. Substitute the new set (9, 12, 15) into the Pythagorean equation $a^2 + b^2 = c^2$ and solve to check.

$$9^2 + 12^2 = 15^2$$
$$81 + 144 = 225$$
$$225 = 225$$

BRAIN TICKLERS   Set # 26

Which of the following sets of three numbers are Pythagorean triples?

1. (5, 12, 13)

2. (6, 7, 8)

3. (7, 24, 25)

4. (11, 60, 61)

(Answers are on page 94.)

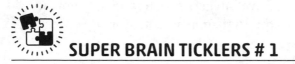

SUPER BRAIN TICKLERS # 1

1. Is the set of three numbers (10, 20, 30) a Pythagorean triple?

2. Is the set of three numbers (18, 80, 82) a Pythagorean triple?

(Answers are on page 94.)

30-60-90-Degree Triangles

Right triangles have one right angle and two acute angles. The sum of the measures of both acute angles of a right triangle is 90 degrees. Triangles with angles of 30-60-90 degrees have unique properties.

Formula: The ratio of the lengths of the sides of the 30-60-90-degree triangle is always $1:\sqrt{3}:2$.

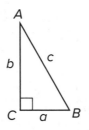

Given triangle ABC where angle $A = 30$ degrees, angle $B = 60$ degrees, and angle $C = 90$ degrees.

Side a is opposite angle A. Side a is the shorter leg of the triangle and has a length of 1.

Side b is opposite angle B. Side b is the longer leg of the triangle and has a length of $\sqrt{3}$.

Side c is opposite angle C. Side c is the hypotenuse of the right triangle and has a length of 2.

Substitute the numbers $1, \sqrt{3}, 2$ into the Pythagorean formula to check this ratio.

$$a^2 + b^2 = c^2$$
$$1^2 + \left(\sqrt{3}\right)^2 = 2^2$$
$$1 + 3 = 4$$

The length of the sides of a 30-60-90-degree triangle is always a multiple of the ratio $1:\sqrt{3}:2$.

The length of the sides of a 30-60-90-degree triangle can be expressed as $1x:\sqrt{3}x:2x$.

Example:

If side a is 2, the lengths of the sides of a 30-60-90-degree triangle would be $2:2\sqrt{3}:4$.
If side a is $\sqrt{3}$, the lengths of the sides of a 30-60-90-degree triangle would be $\sqrt{3}:3:2\sqrt{3}$.
If side a is 10, the lengths of the sides of a 30-60-90-degree triangle would be $10:10\sqrt{3}:20$.

PAINLESS TIP

The length of the hypotenuse of a 30-60-90-degree triangle is always double the length of the shortest side.

If you know the length of one side of a 30-60-90-degree triangle, it's painless to find the length of the other two sides.

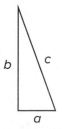

Hint 1: If you know the length of side a, use the following formulas to find the lengths of sides b and c.

$$b = a\sqrt{3}$$
$$c = 2a$$

Example:

If the length of side a of a 30-60-90-degree triangle is 7, what are the lengths of the other two sides?

Step 1: Solve for b. The length of side $b = a\sqrt{3}$. $b = 7\sqrt{3}$

Step 2: Solve for c. The length of side c is equal to twice the length of side a. $c = 2(7) = 14$

The sides of the triangle are $7 : 7\sqrt{3} : 14$.

Hint 2: If you know the length of side b, use the following formulas to find the lengths of sides a and c.

$$a = \frac{b}{\sqrt{3}} = b\frac{\sqrt{3}}{3}$$
$$c = 2a$$

Example:

If the length of side b of the triangle is $4\sqrt{3}$, what are the lengths of the other two sides?

Step 1: Solve for a. The length of side $a = b\dfrac{\sqrt{3}}{3}$.

The length of side b is $4\sqrt{3}$. Create an equation.

$$a = \left(4\sqrt{3}\right)\left(\frac{\sqrt{3}}{3}\right) = \frac{\left(4\sqrt{3}\sqrt{3}\right)}{3}$$

$$a = \frac{4(3)}{3}$$

$$a = 4$$

Step 2: Solve for c. The length of side $c = 2a$.

Since $a = 4$, $c = 2(4)$

$$c = 8$$

The sides of the triangle are $4 : 4\sqrt{3} : 8$.

Hint 3: If you know the length of side c, use the following formulas to find the lengths of sides a and b.

$$a = \frac{c}{2}$$

$$b = a\sqrt{3}$$

Example:

If side c (the hypotenuse) of a 30-60-90-degree triangle is 20, what are the lengths of the two legs?

Step 1: Solve for a. The length of side $a = \frac{c}{2}$.

$$a = \frac{20}{2}$$

$$a = 10$$

Step 2: Solve for b. The length of side $b = a\sqrt{3}$.

$$b = 10\sqrt{3}$$

The sides of the triangle are $10 : 10\sqrt{3} : 20$.

BRAIN TICKLERS   Set # 27

1. If a triangle has sides 3, 4, 5, is it a 30-60-90-degree triangle?

2. If a triangle has sides 50, $50\sqrt{3}$, 100, is it a 30-60-90-degree triangle?

3. If a triangle has sides $\sqrt{3}$, 3, $2\sqrt{3}$, is it a 30-60-90-degree triangle?

4. The length of side a of a 30-60-90-degree triangle is 9. What are the lengths of sides b and c?

5. The length of side b of a 30-60-90-degree triangle is $6\sqrt{3}$. What are the lengths of sides a and c?

6. The length of side c of a 30-60-90-degree triangle is 9. What are the lengths of sides a and b?

(Answers are on page 94.)

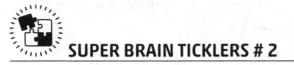

SUPER BRAIN TICKLERS # 2

1. If the shortest leg of a 30-60-90-degree triangle has a length of 10, what is the length of the other leg and the hypotenuse?

2. If the hypotenuse of a 30-60-90-degree triangle has a length of 10, what are the lengths of the legs?

(Answers are on page 94.)

45-45-90-Degree Triangles

A triangle whose angles measure 45, 45, and 90 degrees is an isosceles right triangle. It has two equal angles, two equal sides, and one right angle.

THEOREM   The ratio of the lengths of the sides of every isosceles right triangle is 1:1:$\sqrt{2}$.

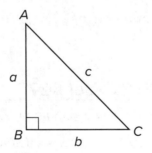

If you know the length of one of the sides of an isosceles right triangle, you can figure out the lengths of the other two sides using the following formulas:

$$\text{If } a = x, b = x, \text{ and } c = x\sqrt{2}.$$

Example:

If the length of side *a* of an isosceles right triangle is 4, what are the lengths of the other two sides?

Step 1: Find the length of side b. If $a = 4$, then $b = 4$ because the two legs of the triangle are equal.

Step 2: Find the length of side c. $c = a\sqrt{2}$, $c = 4\sqrt{2}$

The sides of the triangle ABC are $4, 4, 4\sqrt{2}$.

If you know the length of the hypotenuse of an isosceles right triangle, you can figure out the length of the other two sides using the following formulas:

Step 1: Find the length of side a. $a = \dfrac{c\sqrt{2}}{2}$

Step 2: Find the length of side b. $b = \dfrac{c\sqrt{2}}{2}$

Example:

If the length of the hypotenuse c of an isosceles right triangle is 10, what are the lengths of the legs a and b?

Step 1: Find the length of side a. $a = \dfrac{c\sqrt{2}}{2}$

$$a = \dfrac{10\sqrt{2}}{2}$$

$$a = 52\sqrt{2}$$

Step 2: Find the length of side b. $b = \dfrac{c\sqrt{2}}{2}$

$$b = 5\sqrt{2}$$

If the hypotenuse of an isosceles right triangle is 10, the legs are each $5\sqrt{2}$.

CAUTION—Major Mistake Territory!

The sum of the acute angles of any right triangle is 90 degrees. The acute angles of all isosceles right triangles measure 45 degrees each. You cannot have an isosceles right triangle with angles measuring 30-30-90 degrees because the sum of the three angles is 150 degrees.

BRAIN TICKLERS   Set # 28

1. If each of the legs of an isosceles right triangle has a length of 2, what is the length of the hypotenuse?

2. If each of the legs of an isosceles right triangle has a length of $\frac{1}{2}$, what is the length of the hypotenuse?

3. If each of the legs of an isosceles right triangle has a length of $4\sqrt{2}$, what is the length of the hypotenuse?

4. If the hypotenuse of an isosceles right triangle has a length of $3\sqrt{2}$, what are the lengths of each of the legs?

5. If the hypotenuse of an isosceles right triangle has a length of 1, what are the lengths of each of the legs?

6. If the hypotenuse of an isosceles right triangle has a length of 25, what are the lengths of each of the legs?

(Answers are on page 95.)

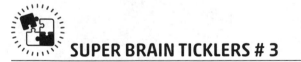

SUPER BRAIN TICKLERS # 3

1. If the legs of an isosceles triangle have a length of 5, what is the length of the hypotenuse?

2. If the hypotenuse of an isosceles right triangle has a length of $100\sqrt{2}$, what are the lengths of each of the legs?

(Answers are on page 95.)

BRAIN TICKLERS   Set # 29

1. If a right triangle has legs 2 inches long and 3 inches long, what is the square of the hypotenuse?

2. If a right triangle has sides measuring 9 inches and 12 inches, what is the length of the hypotenuse?

3. If a triangle has one side of 3 inches and a 5-inch hypotenuse, what is the length of the third side?

(Answers are on page 95.)

SUPER BRAIN TICKLERS # 4

1. What are the measures of the angles of an isosceles right triangle?

2. What type of triangle has angles with measures 30-50-100?

3. What are the measures of the angles of an equilateral triangle?

4. What is the area of an isosceles right triangle with one leg 10 inches long?

(Answers are on page 95.)

BRAIN TICKLERS—THE ANSWERS

Set # 15, page 62

1. 30 degrees

2. 10 degrees

3. 80 degrees

Set # 16, page 66

1. 95 degrees

2. 155 degrees

3. 25 degrees

4. 155 degrees

5. 85 degrees

6. 95 degrees

7. 85 degrees

8. 120 degrees

9. 120 degrees

10. 60 degrees

Set # 17, page 68

1. True

2. False

3. False

Set # 18, page 68

1. True

2. True

3. False

Set # 19, page 70

1. False

2. True

3. True

Set # 20, page 72

1. False

2. False

3. True

Set # 21, page 73

1. True

2. True

3. False

Set # 22, page 75

1. True

2. True

3. False

4. True

Set # 23, page 75

1. R, I

2. S, A

3. R, S

4. E, A

5. I, A

6. O, S

7. I, O

8. O, S

Set # 24, page 77

1. 26

2. 15

3. $4 + \sqrt{8}$

Set # 25, page 80

1. 12 square inches

2. 20 square inches

Set # 26, page 83

1. Yes

2. No

3. Yes

4. Yes

Super Brain Ticklers # 1, page 84

1. No

2. Yes

Set # 27, page 87

1. No

2. Yes

3. Yes

4. $b = 9\sqrt{3}, c = 18$

5. $a = 6, c = 12$

6. $a = 4.5, b = 4.5\sqrt{3}$

Super Brain Ticklers # 2, page 88

1. $a = 10, b = 10\sqrt{3}, c = 20$

2. $a = 5, b = 5\sqrt{3}, c = 10$

Set # 28, page 90

1. $2\sqrt{2}$
2. $\dfrac{\sqrt{2}}{2}$
3. 8
4. 3
5. $\dfrac{\sqrt{2}}{2}$
6. $\dfrac{25\sqrt{2}}{2}$

Super Brain Ticklers # 3, page 90

1. $c = 5\sqrt{2}$
2. $a = 100, b = 100$

Set # 29, page 90

1. 13 square inches

2. 15 inches

3. 4 inches

Super Brain Ticklers # 4, page 91

1. 45-45-90

2. Obtuse and scalene

3. 60-60-60

4. 50 square inches

Chapter 5

Similar and Congruent Triangles

Triangle Parts

Every triangle has three angles and three sides. There are special ways to refer to the angles and sides of a triangle and their relationships between each other. The words *adjacent*, *opposite*, and *included* are used to refer to the relationships between the sides and angles of a triangle.

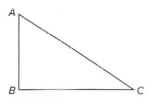

$\overline{AB}$ and $\overline{AC}$ are both adjacent to $\angle A$.
$\overline{BC}$ and $\overline{BA}$ are both adjacent to $\angle B$.
$\overline{CA}$ and $\overline{CB}$ are both adjacent to $\angle C$.

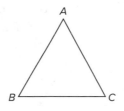

$\overline{AB}$ is included between $\angle A$ and $\angle B$.
$\overline{AC}$ is included between $\angle A$ and $\angle C$.
$\overline{BC}$ is included between $\angle B$ and $\angle C$.

$\angle A$ is included between $\overline{AB}$ and $\overline{AC}$.
$\angle B$ is included between $\overline{AB}$ and $\overline{BC}$.
$\angle C$ is included between $\overline{BC}$ and $\overline{AC}$.

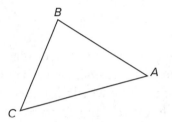

$\overline{BC}$ is opposite angle $\angle A$.
$\overline{AB}$ is opposite angle $\angle C$.
$\overline{AC}$ is opposite angle $\angle B$.

Triangle Midpoints

__MIDPOINT THEOREM__   The segment joining the midpoints of two sides of a triangle is parallel to the third side and half as long as the third side.

In triangle ABC, the midpoints of $\overline{AB}, \overline{AC}$, and $\overline{BC}$ are the points D, E, and F, respectively. DE connects the midpoints of $\overline{AB}$ and $\overline{AC}$ and is parallel to $\overline{BC}$. $\overline{DE}$ is half as long as $\overline{BC}$. $\overline{DE} = \overline{BF} = \overline{FC}$.

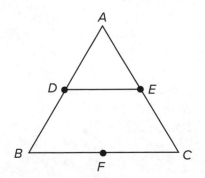

$\overline{DF}$ connects the midpoints of $\overline{AB}$ and $\overline{BC}$ and is half as long as $\overline{AC}$. $\overline{DF}$ is also parallel to $\overline{AC}$. $\overline{DF} = \overline{AE} = \overline{EC}$.

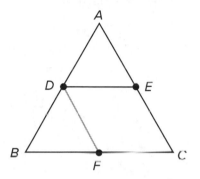

$\overline{EF}$ connects $\overline{AC}$ and $\overline{BC}$ and is half as long as $\overline{AB}$. $\overline{EF}$ is also parallel to $\overline{AB}$, $\overline{EF} = \overline{AD} = \overline{DB}$.

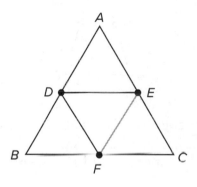

Example:

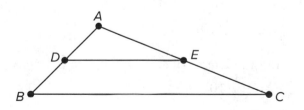

If D is the midpoint of $\overline{AB}$ and E is the midpoint of $\overline{AC}$, then $\overline{DE}$ is parallel to $\overline{BC}$, and the length of $\overline{DE}$ is half the length of $\overline{BC}$. If $\overline{DE} = 3$, what is the length of $\overline{BC}$?

$$\overline{DE} = \frac{1}{2}\overline{BC}$$

$$3 = \frac{1}{2}\overline{BC}$$

$$2(3) = 2\left(\frac{1}{2}\overline{BC}\right)$$

$$6 = \overline{BC}$$

Example:

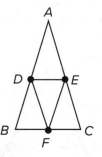

In an isosceles triangle with sides of 10 and a base of 4, what are the lengths of $\overline{DE}, \overline{EF}$, and $\overline{DF}$? Assume D is the midpoint of $\overline{AB}, E$ is the midpoint of $\overline{AC}$, and F is the midpoint of $\overline{BC}$.

Step 1: Find the length of side $\overline{DE}$. Since $\overline{BC} = 4, \overline{DE} = \frac{1}{2}\overline{BC} = \frac{1}{2}(4) = 2$.

Step 2: Find the length of side $\overline{DF}$. Since $\overline{AC} = 10, \overline{DF} = \frac{1}{2}\overline{AC} = \frac{1}{2}(10) = 5$.

Step 3: Find the length of sides $\overline{EF}$. Since $\overline{AB} = 10, \overline{EF} = \frac{1}{2}\overline{AB} = \frac{1}{2}(10) = 5$.

The result is an isosceles triangle DEF with sides $5, 5, 2$.

BRAIN TICKLERS   Set # 30

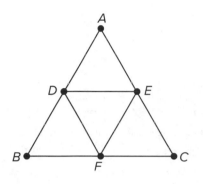

Triangle *ABC* is an equilateral triangle. All the sides are equal. All the angles are equal. Points *D*, *E*, and *F* are midpoints. If $\overline{AB} = 10$, what are the lengths of the following segments?

1. $\overline{BC}$

2. $\overline{EF}$

3. $\overline{AC}$

4. $\overline{AF}$

5. What type of triangle is triangle *DEF*?

6. What is the measure of angle *DEF*?

(Answers are on page 119.)

SUPER BRAIN TICKLERS # 1

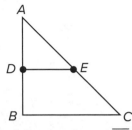

If triangle *ABC* is an isosceles right triangle and $\overline{DE}$ is 4 inches, what are the lengths of the following segments? Assume *D* is the midpoint of $\overline{AB}$ and *E* is the midpoint of $\overline{AC}$.

1. $\overline{BC}$

2. $\overline{AD}$

3. $\overline{DB}$

4. $\overline{AB}$

5. $\overline{AC}$

(Answers are on page 119.)

BRAIN TICKLERS   Set # 31

Look at triangle *XYZ*.

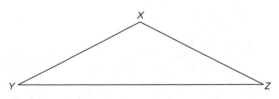

1. What angle is opposite $\overline{XZ}$?

2. What side is opposite ∠*YZX*?

3. What angle is included between $\overline{YX}$ and $\overline{XZ}$?

4. What side is included between ∠*Y* and ∠*Z*?

(Answers are on page 119.)

Similar Triangles

Look at these two triangles.

Triangle *ABC*   Triangle *DEF*

m∠*ABC* = m∠*DEF*
m∠*BCA* = m∠*EFD*
m∠*CAB* = m∠*FDE*

but

$\overline{AB}$ is *not equal* to $\overline{DE}$.
$\overline{BC}$ is *not equal* to $\overline{EF}$.
$\overline{CA}$ is *not equal* to $\overline{FD}$.

These two triangles are not congruent, but they are similar.

- Similar triangles have equal angles, but they do not have equal sides.
- If two triangles are similar, one looks like a larger or smaller version of the other.
- If two triangles are similar, the sides of one are in direct proportion to the sides of the other.
- If two triangles are similar, the ratio of corresponding sides of one triangle is equal to the ratio of corresponding sides of another triangle.

You can prove two triangles are similar using the *AA Similarity Postulate.*

ANGLE-ANGLE (AA) SIMILARITY POSTULATE   If two angles of one triangle are congruent to two angles of another triangle, the triangles are similar.

Look at these two similar triangles.

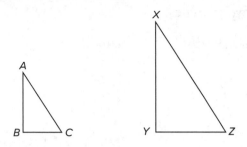

If these two triangles are similar, then

$$\frac{\overline{AB}}{\overline{XY}} = \frac{\overline{BC}}{\overline{YZ}} = \frac{\overline{AC}}{\overline{XZ}}$$

You can use the ratios to find the length of missing sides of a triangle.

Example:

These two triangles are similar.

If $\overline{AB} = 2$
$\overline{BC} = 3$
$\overline{AC} = 5$
$\overline{DE} = 8$

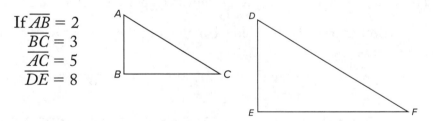

Find the length of $\overline{EF}$ and $\overline{DF}$.

First match up corresponding sides.

$$\frac{\overline{AB}}{\overline{DE}} = \frac{\overline{BC}}{\overline{EF}} = \frac{\overline{AC}}{\overline{DF}}$$

Now substitute the sides you know.

$$\frac{\overline{AB}}{\overline{DE}} = \frac{\overline{BC}}{\overline{EF}} = \frac{\overline{AC}}{\overline{DF}}$$

Take one pair of ratios.

$$\frac{2}{8} = \frac{3}{\overline{EF}}$$

Cross-multiply to solve.

$$2(\overline{EF}) = 3(8)$$
$$2(\overline{EF}) = 24$$
$$\overline{EF} = 12$$

Now solve for $\overline{DF}$.

$$\frac{2}{8} = \frac{5}{\overline{DF}}$$

$$2(\overline{DF}) = 5(8)$$
$$2(\overline{DF}) = 40$$
$$\overline{DF} = 20$$

 **BRAIN TICKLERS   Set # 32**

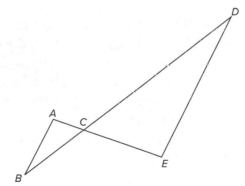

If triangle ABC and triangle ECD are similar and if $\overline{AB} = 2$, $\overline{AC} = 1$, $\overline{DE} = 5$, and $\overline{CD} = 6$, find $\overline{BC}$ and $\overline{CE}$.

(Answers are on page 119.)

Right Triangle Proportions Theorem

Triangle ABC is a right triangle. Angle BAC is a right angle. Angle ABC and angle ACB are acute angles. $\overline{AD}$ is the altitude of the triangle. The hypotenuse $\overline{BC}$ is now divided into two segments, $\overline{BD}$ and $\overline{DC}$. The altitude of a right triangle is a line segment from the vertex of the right angle perpendicular to the hypotenuse.

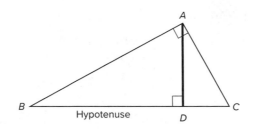

Now there are three triangles: a large right triangle, triangle ABC; a medium right triangle, triangle ABD; and a small right triangle, triangle ADC.

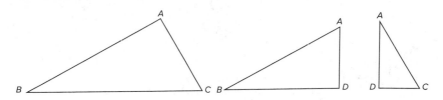

The three triangles are similar but not congruent to each other. Triangle ABC, triangle ABD, and triangle ADC have the same shape but they are not the same size. Their corresponding angles are equal. The right angles are equal: angle BAC = angle ADB = angle ADC. The smaller of the two acute angles in each of the triangles are also equal: angle ABC = angle ABD = angle CAD. The larger of the two acute angles in each of the triangles are also equal: angle ACB = angle BAD = angle ACD.

THEOREM   If an altitude is drawn from the right angle of any right triangle, then the two triangles formed are similar to the original triangle and similar to each other.

The small triangle (ACD) and the large triangle (ABC) are similar to each other. The short leg of the small triangle divided by the hypotenuse of the small triangle = the short leg of the large triangle divided by the hypotenuse of the large triangle. Notice that segment AC is both the hypotenuse of the small triangle and the short leg of the large triangle.

$$\frac{\overline{CD}}{\overline{AC}} = \frac{\overline{AC}}{\overline{BC}}$$

The medium triangle (ABD) and the large triangle (ABC) are similar to each other. The large leg of the medium triangle divided by the hypotenuse of the medium triangle = the large leg of the large triangle divided by the hypotenuse of the large triangle. Notice that segment AB is both the hypotenuse of the medium triangle and the large leg of the large triangle.

$$\frac{\overline{BD}}{\overline{AB}} = \frac{\overline{AB}}{\overline{BC}}$$

The small triangle ACD and the medium triangle ABD are similar to each other. The small leg of the small triangle divided by the large leg of the small triangle = the small leg of the medium triangle divided by the large leg of the medium triangle. Notice segment AD is both the large leg of the small triangle and the small leg of the medium triangle.

$$\frac{\overline{DC}}{\overline{AD}} = \frac{\overline{AD}}{\overline{BD}}$$

Understanding these ratios makes it possible to solve ratio problems.

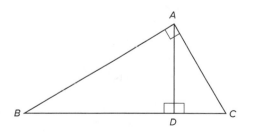

Example:

Triangle *ABC* is a right triangle. Angle *BAC* is a right angle. Segment *BC* is the hypotenuse of the triangle. Segment *AD* is the altitude of triangle *ABC*. Angle *ADB* and angle *ADC* are right angles. Assume side *DC* = 5 and side *BC* = 20. Find the lengths of sides *AB* and *AC*.

Step 1: Find $\overline{BD}$.
$$\overline{BD} = \overline{BC} - \overline{DC}$$
$$20 - 5 = 15$$

Step 2: Use the proportions to find $\overline{AC}$.
$$\frac{\overline{DC}}{\overline{AC}} = \frac{\overline{AC}}{\overline{BC}}$$
$$\frac{5}{x} = \frac{x}{20}$$
$$x^2 = 100$$
$$x = 10$$
$$\overline{AC} = 10$$

Step 3: Use the proportions to find $\overline{AB}$.
$$\frac{\overline{BD}}{\overline{AB}} = \frac{\overline{AB}}{\overline{BC}}$$
$$\frac{15}{x} = \frac{x}{20}$$
$$x^2 = 300$$
$$\overline{AB} = 10\sqrt{3}$$

BRAIN TICKLERS   Set # 33

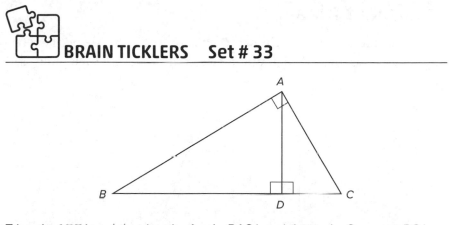

Triangle *ABC* is a right triangle. Angle *BAC* is a right angle. Segment *BC* is the hypotenuse of triangle *ABC*. Segment *AD* is the altitude of triangle *ABC*. Angle *ADB* and angle *ADC* are right angles. Assume $\overline{BC}$ = 12 and $\overline{BD}$ = 9.

1. What is the measure of $\overline{CD}$?

2. What is the measure of $\overline{AC}$?

(Answers are on page 119.)

SUPER BRAIN TICKLERS #2

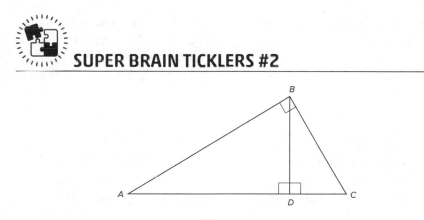

Triangle *ABC* is a right triangle. $\overline{AC}$ is the hypotenuse of triangle *ABC*. $\overline{BD}$ is the altitude.

Assume $\overline{AD}$ = 16 and $\overline{AC}$ = 20.

1. Find $\overline{DC}$.   3. Find $\overline{AB}$.

2. Find $\overline{BD}$.   4. Find $\overline{BC}$.

(Answers are on page 120.)

Congruent Triangles

Congruent triangles have the same size and the same shape. Two triangles are congruent if their angles are congruent *and* their sides are congruent.

These two triangles are congruent.

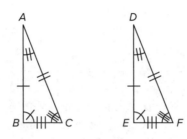

$$\angle A \cong \angle D$$
$$\angle B \cong \angle E$$
$$\angle C \cong \angle F$$
$$\overline{AB} \cong \overline{DE}$$
$$\overline{BC} \cong \overline{EF}$$
$$\overline{CA} \cong \overline{FD}$$

In order for two triangles to be congruent, all six of these statements must be true.

If two triangles are congruent, then their corresponding parts are congruent.

Example:

If triangle *ABC* is congruent to triangle *DEF* and the measure of angle *ABC* is 90 degrees, then the measure of angle *DEF* is also 90 degrees.

If triangle *ABC* is congruent to triangle *DEF* and side $AB = 4$, then side $DE = 4$.

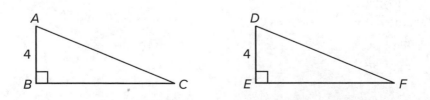

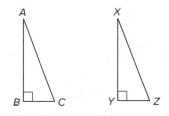

BRAIN TICKLERS   Set # 34

Triangle *ABC* is congruent to triangle *XYZ*.

The measure of angle $B = 90$, the measure of angle $A = 20$, and $\overline{XY} = 2$.

  1. What is the m∠*X*?

  2. What is the m∠*Y*?

  3. What is the m∠*Z*?

  4. What is the length of $\overline{AB}$?

(Answers are on page 120.)

Corresponding angles and sides

When two triangles are congruent, all pairs of corresponding sides and all pairs of corresponding angles are congruent. Corresponding sides and angles are the sides and angles that match up between two triangles. Mathematicians use slash marks to indicate corresponding sides or angles.

Example:

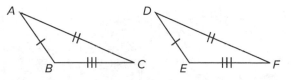

The single slash mark means $\overline{AB}$ is congruent to $\overline{DE}$. The double slash mark means $\overline{AC}$ is congruent to $\overline{DF}$. The triple slash mark means $\overline{BC}$ is congruent to $\overline{EF}$.

SSS (SIDE-SIDE-SIDE) POSTULATE   If three sides of one triangle are congruent to three sides of another triangle, then the triangles are congruent.

Example:

To prove triangle $ABC \cong$ triangle XYZ, show that $AB \cong XY$, $AC \cong XZ$, and $BC \cong YZ$.

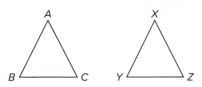

MINI-PROOF

Figure *ABDC* is a rectangle.
Show that triangle *ABD* is congruent to triangle *ACD*.

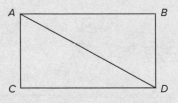

To prove these two triangles congruent, show that all three sides are equal to each other. In other words, $\overline{AC} \cong \overline{BD}$, $\overline{AB} \cong \overline{CD}$, and $\overline{AD} \cong \overline{AD}$.

First list what you know.

1. Figure *ABDC* is a rectangle.

2. $\overline{AD}$ is a diagonal of rectangle *ABDC*.

Next list what you can infer based on what you know.

3. $\overline{AC} \cong \overline{BD}$ since they are opposite sides of a rectangle, and opposite sides of a rectangle are always congruent.

4. $\overline{AB} \cong \overline{CD}$ since they are opposite sides of a rectangle, and opposite sides of a rectangle are always congruent.

5. $\overline{AD} \cong \overline{AD}$ since every line segment is congruent to itself.

Conclusion
Since all three sides of one triangle are congruent to all three sides of the other triangle, the triangles are congruent. Since the triangles are congruent,
$\angle ABD \cong \angle ACD$
$\angle DAB \cong \angle ADC$
$\angle BDA \cong \angle CAD$

SAS (SIDE-ANGLE-SIDE) POSTULATE If two sides and the included angle of one triangle are congruent to two sides and the included angle of a second triangle, then the triangles are congruent.

Example:

To prove $\triangle ABC = \triangle XYZ$, show that $\overline{AB} \cong \overline{XY}$, $\overline{AC} \cong \overline{XZ}$, and $\angle A$ is congruent to $\angle X$. $\angle A$ is included between $\overline{AB}$ and $\overline{AC}$, and $\angle X$ is included between $\overline{XY}$ and $\overline{XZ}$. You can also prove that $\triangle ABC$ is congruent to $\triangle XYZ$ using $\angle B$ and $\angle Y$ and their adjacent sides, or $\angle C$ and $\angle Z$ and their adjacent sides.

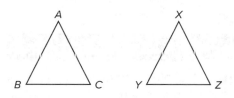

MINI-PROOF

Figure *WXZY* is a rhombus.

$\overline{WZ}$ and $\overline{XY}$ are diagonals of the rhombus. Is $\triangle WEY$ congruent to $\triangle ZEX$?

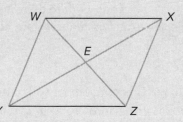

To prove triangles *WEY* and *ZEX* congruent using SAS, show that two sides and an included angle of one triangle are congruent to two sides and an included angle of another triangle.

First list what you know.

1. Figure *WXZY* is a rhombus.

2. $\overline{WZ}$ is a diagonal of the rhombus.

3. $\overline{XY}$ is a diagonal of the rhombus.

Next list what you can infer based on what you know.

4. The m∠*WEY* = m∠*XEZ* since they are vertical angles, and all vertical angles are equal. Since the measures are equal, the angles are congruent.

5. $\overline{WZ}$ bisects $\overline{XY}$ since the diagonals of a rhombus bisect each other.

6. $\overline{WE} = \overline{EZ}$ since a bisected line segment is divided into two congruent segments.

7. $\overline{XE} = \overline{EY}$ since a bisected line segment is divided into two congruent segments.

Conclusion

$\triangle WEY$ is congruent to $\triangle ZEX$ since two sides and the included angle of one triangle are congruent to two sides and the included angles of another triangle.

CAUTION—Major Mistake Territory!

You *cannot* prove two triangles congruent if two sides of one triangle are congruent to two sides of another triangle and an angle that is not included is congruent to an angle that is not included. *The angle must be included for the two triangles to be congruent.*

ASA (ANGLE-SIDE-ANGLE) POSTULATE   If two angles and the included side of one triangle are congruent to two angles and the included side of another triangle, then the two triangles are congruent.

Example:

To prove triangle ABC is congruent to triangle XYZ you must show that angle A is congruent to angle X, angle B is congruent to angle Y, and the included side AB is congruent to the included side XY.

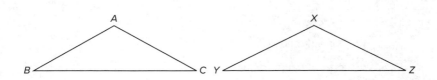

MINI-PROOF

Figure ABDC is a square. $\overline{AD}$ and $\overline{BC}$ are diagonals of the square.

Prove triangle *ADC* is congruent to triangle *BCD*.

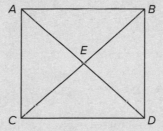

First list what you know.

1. *ABDC* is a square.

2. $\overline{AD}$ is a diagonal of square *ABDC*.

3. $\overline{BC}$ is a diagonal of square *ABDC*.

Next list what you can infer from what you know.

4. $\overline{CD} = \overline{CD}$ since every segment is congruent to itself.

5. $\angle C \cong \angle D$ since they are both right angles.

6. $\overline{BC} = \overline{AD}$ since the diagonals of a square are congruent.

7. $\overline{EC} = \overline{ED}$ since the diagonals of a square bisect each other.

8. $\triangle ECD$ is an isosceles triangle.

9. $\angle ECD \cong \angle EDC$, since the base angles of an isosceles triangle are congruent.

Conclusion

Using the ASA postulate, $\triangle ADC$ and $\triangle BCD$ are congruent since two angles and the included side of each triangle are congruent. You could also prove these two triangles congruent using SAS.

THEOREM   Two triangles are congruent if two angles and the side opposite one of them are congruent.

Example:

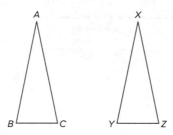

To prove $\triangle ABC \cong \triangle XYZ$:

- show that $\angle A \cong \angle X$ and $\angle B \cong \angle Y$. Now show that $\overline{BC} \cong \overline{YZ}$ or $\overline{AC} \cong \overline{XZ}$, since these are all nonincluded sides.

or
- show that $\angle B \cong \angle Y$ and $\angle C \cong \angle Z$ and either $\overline{AB} \cong \overline{XY}$ or $\overline{AC} \cong \overline{XZ}$, since these are all nonincluded sides.

or
- show that $\angle A \cong \angle X$ and $\angle C \cong \angle Z$ and $\overline{AB} \cong \overline{XY}$ or $\overline{BC} \cong \overline{YZ}$, since these are all nonincluded sides.

MINI-PROOF

ABDC is a parallelogram. A parallelogram is a four-sided figure in which opposite sides are parallel and equal. $\overline{AD}$ is perpendicular to $\overline{CD}$ and $\overline{AB}$.

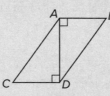

Prove that $\triangle ACD$ is congruent to $\triangle DBA$.

First list what you know.

1. *ABDC* is a parallelogram.

2. $\overline{AD}$ is perpendicular to $\overline{CD}$ and $\overline{AB}$.

What does this mean?

3. $\angle B \cong \angle C$, since the opposite angles of a parallelogram are congruent.

4. $\overline{AD} \cong \overline{AD}$, since every segment is congruent to itself.

5. ∠ADC and ∠DAB are right angles.

6. m∠ADC = m∠DAB, since all right angles are 90 degrees.

Conclusion

△ACD is congruent to △DBA, since two angles and a nonincluded side of one triangle are equal to two angles and a nonincluded side of another triangle.

HYPOTENUSE-LEG POSTULATE   Two right triangles are congruent if the hypotenuse and leg of one right triangle are congruent to the hypotenuse and leg of another right triangle.

To prove that these two right triangles are congruent, just show that hypotenuse $\overline{AC}$ is congruent to hypotenuse $\overline{XZ}$ and leg $\overline{AB}$ is congruent to leg $\overline{XY}$. Or show that hypotenuse $\overline{AC}$ is congruent to hypotenuse $\overline{XZ}$ and leg $\overline{BC}$ is congruent to leg $\overline{YZ}$.

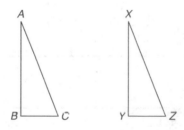

MINI-PROOF

Triangle ABC is an isosceles triangle where $\overline{AB} \cong \overline{AC}$.

$\overline{AD}$ is the perpendicular bisector of $\overline{BC}$. Prove △ABD ≅ △ACD.

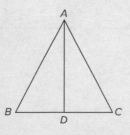

First list what you know.

1. Triangle *ABC* is an isosceles triangle.

2. $\overline{AD}$ is the altitude of $\triangle ABC$.

What does this mean?

3. $\overline{AD}$ is perpendicular to $\overline{BC}$ since $\overline{AD}$ is an altitude of triangle *ABC*.

4. Angle *ADB* and angle *ADC* are both right angles since $\overline{AD}$ is perpendicular to $\overline{BC}$.

5. Triangle *ADB* and triangle *ADC* are both right triangles, since angle *ADB* and angle *ADC* are both right angles.

6. The hypotenuse of triangle *ABD* is congruent to the hypotenuse of triangle *ACD*, since triangle *ABC* is an isosceles triangle.

7. One leg of triangle *ABD* is congruent to one leg of triangle *ACD*, since $\overline{AD}$ is a leg of both triangles and congruent to itself.

Conclusion

Triangle *ABD* is congruent to triangle *ACD* because of the hypotenuse-leg postulate. The hypotenuse and one leg of triangle *ABD* are congruent to the hypotenuse and one leg of triangle *ACD*.

PAINLESS TIP

There are five ways to prove triangles congruent.
Write each of these five ways on one side of an index card.

SSS	Side, side, side
SAS	Side, angle, side
ASA	Angle, side, angle
AAS	Angle, angle, side
H-L	Hypotenuse-leg

BRAIN TICKLERS—THE ANSWERS

Set # 30, page 101

1. $\overline{BC} = 10$
2. $\overline{EF} = 5$
3. $\overline{AC} = 10$
4. $\overline{AE} = 5$
5. Equilateral
6. 60 degrees

Super Brain Ticklers # 1, page 102

1. 8 inchoo
2. 4 inches
3. 4 inches
4. 8 inches
5. $8\sqrt{2}$ inches

Set # 31, page 102

1. Angle Y, which is also angle XYZ
2. $\overline{XY}$
3. Angle X or angle YXZ
4. $\overline{YZ}$

Set # 32, page 105

1. $\overline{BC} = 2\frac{2}{5}$ and $\overline{CE} = 2\frac{1}{2}$

Set # 33, page 109

1. 3
2. 6

Super Brain Ticklers # 2, page 109

1. $\overline{DC} = 4$
2. $\overline{BD} = 8$
3. $\overline{AB} = 8\sqrt{5}$
4. $\overline{BC} = 4\sqrt{5}$

Set # 34, page 111

1. 20 degrees
2. 90 degrees
3. 70 degrees
4. 2

Chapter 6

Quadrilaterals

Quadrilaterals are four-sided polygons. There are all kinds of quadrilaterals. Squares and rectangles are common quadrilaterals. Trapezoids, parallelograms, and rhombuses are also quadrilaterals. Quadrilaterals are everywhere. A playing card, a window frame, and index cards are all quadrilaterals.

Trapezoids

A *trapezoid* is a quadrilateral with exactly one pair of parallel sides. Both of these figures are trapezoids.

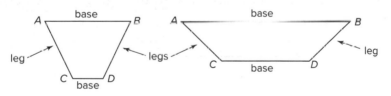

In both figures, $\overline{AB}$ is parallel to $\overline{CD}$. Notice that $\overline{AC}$ is not parallel to $\overline{BD}$.

The parallel sides of the trapezoid are called the *bases* of the trapezoid. $\overline{AB}$ and $\overline{CD}$ are the bases of these trapezoids.

The nonparallel sides of the trapezoid are called the *legs* of the trapezoid. $\overline{AC}$ and $\overline{BD}$ are the legs of these trapezoids.

The *median* of a trapezoid is a line segment that is parallel to both bases of the trapezoid and connects both legs of the trapezoid at their midpoints. In trapezoid $ABCD$, $\overline{EF}$ is the median of the trapezoid. The length of the median of a trapezoid is

$$\frac{1}{2} (\text{base 1} + \text{base 2})$$

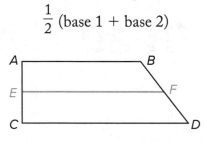

Example:

If one base of a trapezoid is 6 inches, and the other base is 12 inches, what is the length of the median?

$$\text{Length of median} = \frac{1}{2} (6 + 12)$$

$$\text{Length of median} = \frac{1}{2} (18)$$

$$\text{Length of median} = 9 \text{ inches}$$

Here is a picture of a trapezoid. $\overline{EF}$ is the median of the trapezoid.

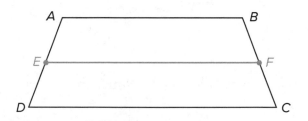

What do you know about this trapezoid?

$\overline{AB}$ and $\overline{CD}$ are the bases of the trapezoid.
$\overline{AD}$ and $\overline{BC}$ are the legs of the trapezoid.
$\overline{AB}$, $\overline{CD}$, and $\overline{EF}$ are all parallel to each other.
$\overline{AE} \cong \overline{ED}$.
$\quad\overline{BF} \cong \overline{FC}$.

Do this experiment to find some common properties of trapezoids.

EXPERIMENT

Explore the properties of trapezoids.

Materials

Paper   Scissors
Pencil   Tape

Procedure

1. On a piece of paper, draw three large triangles. Draw one obtuse triangle, one right triangle, and one isosceles triangle.

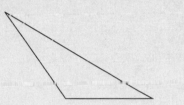 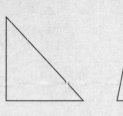

2. Cut out all three of these triangles.

3. Draw a line parallel to the base of each of the triangles.

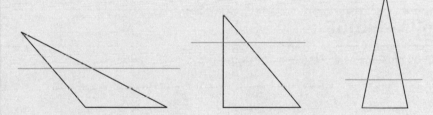

4. Cut off the top of the triangles on the line. The result is three trapezoids.

5. Rip off the angles on the same side of the trapezoid. Tape both of these angles together. Together they should equal 180 degrees.

6. Rip off all four angles from one of the trapezoids. Tape all four of the angles together. What is the sum of all four angles?

Something to think about . . .

If you draw a line on a trapezoid parallel to the base of the trapezoid, two new shapes result. What are they?

If you draw a line down the center of a trapezoid perpendicular to the base of the trapezoid, two new shapes result. What are they?

THEOREM   The two angles on one side of a trapezoid are supplementary.

Angle *A* and angle *C* are supplementary.

$$m\angle A + m\angle C = 180$$

Angle *B* and angle *D* are supplementary.

$$m\angle B + m\angle D = 180$$

THEOREM   The sum of the angles of a trapezoid is 360 degrees.

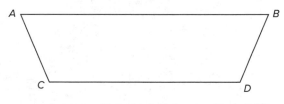

$$m\angle A + m\angle B + m\angle C + m\angle D = 360$$

MINI-PROOF

Prove that the sum of the angles of a trapezoid is 360 degrees.

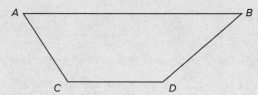

First list what you know.

1. *ABDC* is a trapezoid.

What does this tell you?

2. Angles *A* and *C* are supplementary since both angles on the same side of a trapezoid are supplementary.

3. Angles *B* and *D* are supplementary since both angles on the same side of a trapezoid are supplementary.

4. $m\angle A + m\angle C = 180$ since supplementary angles total 180 degrees.

5. $m\angle B + m\angle D = 180$ since supplementary angles total 180 degrees.

Conclusion

$$m\angle A + m\angle B + m\angle C + m\angle D = 360 \text{ degrees}$$

Isosceles Trapezoids

An *isosceles trapezoid* is a special type of trapezoid, where both legs are congruent. To draw an isosceles trapezoid, start by drawing an isosceles triangle. Draw a line parallel to the base of the triangle. Cut off the top of the triangle along the line you just drew. The result is an isosceles trapezoid.

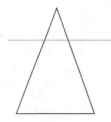

An isosceles trapezoid has two pairs of base angles.

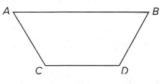

Angles *A* and *B* are one pair of base angles of the trapezoid.

Angles *C* and *D* are the other pair of base angles of the trapezoid.

The base angles of an isosceles trapezoid are congruent.

$\angle A \cong \angle B$.
$\quad \angle C \cong \angle D$.

The legs of an isosceles triangle are congruent.

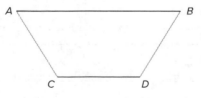

$\overline{AC} \cong \overline{BD}$.

The diagonals of an isosceles trapezoid are congruent.

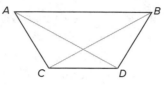

$\overline{AD} \cong \overline{BC}.$

EXPERIMENT

Discover the sum of the angles of an isosceles trapezoid.

Materials
  Paper
  Pencil

Procedure
  1. Draw an isosceles trapezoid.

  2. Draw a single diagonal in the trapezoid. This diagonal divides the trapezoid into two triangles.

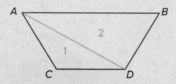

  3. Label the triangles triangle 1 and triangle 2.

  4. What is the measure of the sum of the angles of triangle 1?

  5. What is the measure of the sum of the angles of triangle 2?

  6. Add the sum of the angles of these two triangles together.
$$180 + 180 = 360$$
  The total is 360 degrees.

Something to think about . . .
Is there another way to divide an isosceles trapezoid into two equal shapes?

The sum of the angles of an isosceles trapezoid is 360 degrees.

$$m\angle A + m\angle B + m\angle C + m\angle D = 360$$

THEOREM   The opposite angles of an isosceles trapezoid are supplementary.

Example:

Angles 1 and 4 are supplementary.

$$m\angle 1 + m\angle 4 = 180$$

Angles 2 and 3 are supplementary.

$$m\angle 2 + m\angle 3 = 180$$

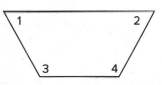

BRAIN TICKLERS   Set # 35

Compute the values of the angles of this isosceles trapezoid.

1. If m∠2 = 30, what is the m∠5?

2. If m∠1 = 30, what is the m∠4?

Compute the measures of the angles for the isosceles trapezoid if m∠3 is 120 degrees.

3. What is the m∠1?

4. What is the m∠2?

5. What is the m∠4?

6. What is the m∠5?

7. What is the m∠6?

(Answers are on page 142.)

Parallelograms

A *parallelogram* is a quadrilateral with two pairs of parallel sides. All these figures are parallelograms.

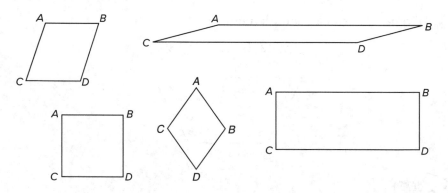

In all these figures, side AB is parallel to side CD, and side AC is parallel to side BD.

THEOREM   Opposite angles of a parallelogram are congruent.

Example:

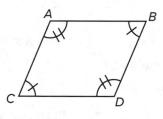

$$\angle A \cong \angle D$$
$$\angle C \cong \angle B$$

THEOREM   The opposite sides of a parallelogram are congruent.

Example:

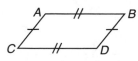

$$AB \cong CD$$
$$AC \cong BD$$

THEOREM   Consecutive pairs of angles of a parallelogram are supplementary.

Example:

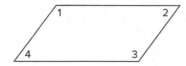

$$m\angle1 + m\angle2 = 180$$
$$m\angle2 + m\angle3 = 180$$
$$m\angle3 + m\angle4 = 180$$
$$m\angle4 + m\angle1 = 180$$

EXPERIMENT

Explore the relationship between the angles of a parallelogram.

Materials
Protractor
Pencil
Paper

Procedure

1. Draw three different parallelograms.

2. Label the angles of each parallelogram angles 1, 2, 3, and 4.

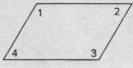

3. Measure each of the angles of the parallelogram. Enter the measurements in the chart.

Parallel-ogram	m∠1	m∠2	m∠3	m∠4	m∠1+ m∠2	m∠1+ m∠4	m∠2+ m∠3	m∠3+ m∠4	m∠1+m∠2+ m∠3+m∠4
1									
2									
3									

4. Add the measurements of the pairs of angles indicated in the chart. Enter the results in the chart.

5. Add the measurements of all four angles together. Enter the results in the chart.

Something to think about . . .

What did you notice about the patterns formed?

Can you draw a parallelogram where the sum of the angles is not 360 degrees?

THEOREM   The sum of the angles of a parallelogram is 360 degrees.

Example:

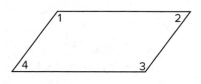

$$m\angle 1 + m\angle 2 + m\angle 3 + m\angle 4 = 360$$

THEOREM   Each diagonal of a parallelogram separates the parallelogram into two congruent triangles.

Example:

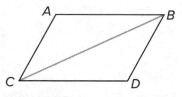

Triangle $ABC \cong$ Triangle BCD

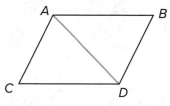

Triangle $ABD \cong$ Triangle ACD

MINI-PROOF

How can you prove triangle *ABC* is congruent to triangle *CAD*?

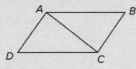

First list what you know.

1. *ABCD* is a parallelogram.

2. $\overline{AC}$ is a diagonal of parallelogram *ABCD*.

What does this tell you?

3. $\overline{AC} \cong \overline{AC}$ since every segment is congruent to itself.

4. $\overline{AB} \cong \overline{CD}$ since opposite sides of a parallelogram are congruent.

5. $\overline{AD} \cong \overline{BC}$ since opposite sides of a parallelogram are congruent.

Conclusion

Triangle *ABC* is congruent to triangle *CAD* since three sides of one triangle are congruent to three sides of another triangle.

THEOREM   The diagonals of a parallelogram bisect each other.

Example:

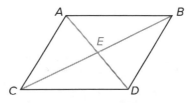

Diagonals *AD* and *BC* intersect at point *E*.

$$\overline{AE} = \overline{ED}$$

$$\overline{CE} = \overline{EB}$$

EXPERIMENT

Compare the diagonals of a parallelogram.

Materials
Pencil
Paper
Scissors

Procedure

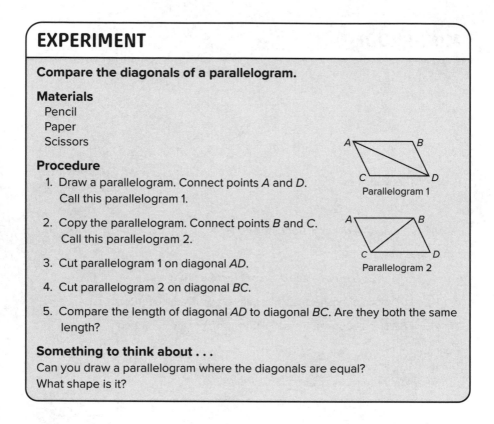

1. Draw a parallelogram. Connect points *A* and *D*.
   Call this parallelogram 1.

Parallelogram 1

2. Copy the parallelogram. Connect points *B* and *C*.
   Call this parallelogram 2.

Parallelogram 2

3. Cut parallelogram 1 on diagonal *AD*.

4. Cut parallelogram 2 on diagonal *BC*.

5. Compare the length of diagonal *AD* to diagonal *BC*. Are they both the same length?

Something to think about . . .
Can you draw a parallelogram where the diagonals are equal?
What shape is it?

BRAIN TICKLERS   Set # 36

This figure is a parallelogram. The m∠ 5 is 110.

1. What is m∠1?

2. What is m∠2?

3. What is m∠3?

4. What is m∠4?

5. What is m∠6?

Find the measure of the indicated angles in this parallelogram when m∠5 = 30, m∠6 = 40, and m∠7 = 50.

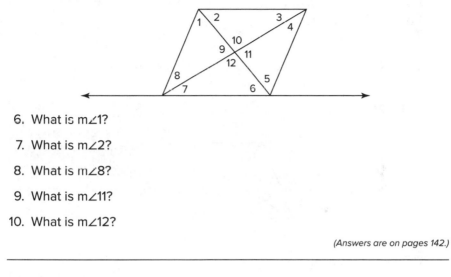

6. What is m∠1?

7. What is m∠2?

8. What is m∠8?

9. What is m∠11?

10. What is m∠12?

(Answers are on pages 142.)

Rhombuses

A *rhombus* is a parallelogram with four equal sides. Since it is a parallelogram, opposite sides of a rhombus are parallel. All these shapes are rhombuses.

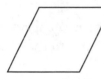 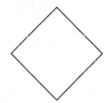

THEOREM   The diagonals of a rhombus bisect the angles of the rhombus.

Example:

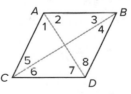

Diagonal $\overline{AD}$ bisects ∠CAB and ∠BDC.

The m∠1 = m∠2.
The m∠7 = m∠8.

Diagonal $\overline{BC}$ bisects $\angle ABD$ and $\angle DCA$.

The m$\angle 3$ = m$\angle 4$.
The m$\angle 5$ = m$\angle 6$.

THEOREM The diagonals of a rhombus are perpendicular.

Example:

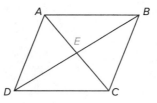

These diagonals intersect at point E.

$\overline{AC}$ is perpendicular to $\overline{BD}$.
$\angle AEB$ is a right angle.
$\angle BEC$ is a right angle.
$\angle CED$ is a right angle.
$\angle DEA$ is a right angle.

Since a rhombus is a special parallelogram with four equal sides, everything that applies to parallelograms also applies to rhombuses.

- The opposite angles of a rhombus are congruent.
- The consecutive pairs of angles of a rhombus are supplementary.
- The diagonals of a rhombus separate the rhombus into two congruent triangles.
- The diagonals of a rhombus bisect each other.

BRAIN TICKLERS   Set # 37

This figure is a rhombus. The measure of ∠5 is 40. Determine the measures of the indicated angles.

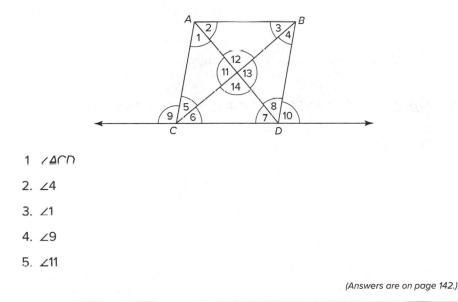

1. ∠ACD

2. ∠4

3. ∠1

4. ∠9

5. ∠11

(Answers are on page 142.)

Rectangles

A *rectangle* is a parallelogram with four right angles. Both figures are rectangles.

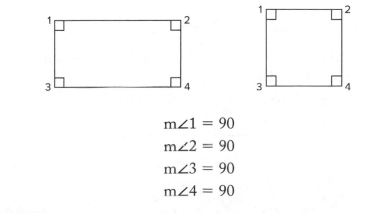

$$m\angle 1 = 90$$
$$m\angle 2 = 90$$
$$m\angle 3 = 90$$
$$m\angle 4 = 90$$

THEOREM   All four angles of a rectangle are congruent.

Example:

Angle A = Angle B = Angle C = Angle D

THEOREM Any two angles of a rectangle are supplementary.

Example:

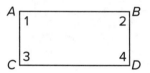

m∠1 + m∠2 = 180 degrees; ∠1 and ∠2 are supplementary.
m∠1 + m∠3 = 180 degrees; ∠1 and ∠3 are supplementary.
m∠1 + m∠4 = 180 degrees; ∠1 and ∠4 are supplementary.
m∠2 + m∠3 = 180 degrees; ∠2 and ∠3 are supplementary.
m∠2 + m∠4 = 180 degrees; ∠2 and ∠4 are supplementary.
m∠3 + m∠4 = 180 degrees; ∠3 and ∠4 are supplementary.

Since a rectangle is a parallelogram, the theorems that apply to parallelograms also apply to rectangles.

- The diagonals of a rectangle separate the rectangle into two congruent triangles.
- The diagonals of a rectangle bisect each other.

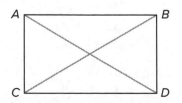

BRAIN TICKLERS Set # 38

This figure is a rectangle, and m∠1 = 30.

1. What is the measure of angle 2?

2. What is the measure of angle 3?

3. What is the measure of angle 4?

4. What is the measure of angle 5?

5. What is the measure of angle 6?

(Answers are on page 143.)

Squares

A *square* is a rectangle with four equal sides.

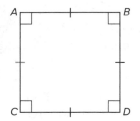

Since a square is a type of rectangle, the measure of all the angles of a square is 90 degrees.

$$m\angle A = 90$$

$$m\angle B = 90$$

$$m\angle C = 90$$

$$m\angle D = 90$$

THEOREM   All four sides of a square are congruent.

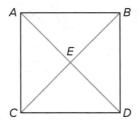

Example:
$$\overline{AB} = \overline{BC} = \overline{CD} = \overline{DA}$$

Look at the relationship between the diagonals of a square.

THEOREM   The diagonals of a square are perpendicular to each other.

Example:

Diagonal $\overline{AD}$ is perpendicular to diagonal $\overline{BC}$.

> Angle AEC is a right angle.
> Angle AEB is a right angle.
> Angle BED is a right angle.
> Angle CED is a right angle.

Since a square is a rectangle and a rectangle is a type of parallelogram, all the theorems that apply to parallelograms and to rectangles also apply to squares.

- The diagonals of squares are equal.
- The measures of all the angles of a square are equal.
- The sum of the angles of a square is 360 degrees.
- The diagonals of a square bisect each other.
- The triangles formed by the diagonals of a square are congruent.

CAUTION—Major Mistake Territory!

Every square is a rectangle, but not every rectangle is a square.

Every square is a parallelogram, but not every parallelogram is a square.

Every square is a quadrilateral, but not every quadrilateral is a square.

Every rectangle is a parallelogram, but not every parallelogram is a rectangle.

Every rectangle is a quadrilateral, but not every quadrilateral is a rectangle.

Every parallelogram is a quadrilateral, but not every quadrilateral is a parallelogram.

EXPERIMENT

Discover the relationship between the four small triangles formed by the diagonals of a square.

Materials
Paper
Pencil
Scissors

Procedure
1. Draw a square.

2. Draw the diagonals of the square.

3. Cut the square along the diagonals and form four small triangles.

4. Place these four triangles on top of each other. Are they congruent?

5. What other shapes can you construct from these four small triangles?

Something to think about . . .
Do the diagonals of a rectangle form four identical triangles?

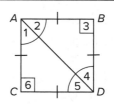

BRAIN TICKLERS   Set # 39

Determine the relationship between the following pairs of angles contained in the square.

C = Complementary angles
S = Supplementary angles
E = Equal angles
V = Vertical angles
A = Adjacent angles

1. ∠CAB and ∠BDC

2. ∠1 and ∠2

3. ∠1 and ∠4

4. ∠3 and ∠6

5. ∠1 and ∠5

(Answers are on page 143.)

QUADRILATERAL SUMMARY CHART

Fill in the following summary chart.
Check the boxes that are true for each quadrilateral.

Features	Trapezoid	Parallelogram	Rhombus	Rectangle	Square
Opposite angles are congruent.					
Opposite angles are supplementary.					
All angles are congruent.					
Diagonals bisect each other.					
Diagonals are perpendicular.					
Opposite sides are congruent.					
Adjacent angles are supplementary.					
Sum of the angles is 360.					

SUPER BRAIN TICKLERS

Decide what type of quadrilateral could be represented by the following statements. Circle all that apply.

 T represents Trapezoid
 P represents Parallelogram
 RH represents Rhombus
 R represents Rectangle
 S represents Square

T P RH R S   1. Two pairs of parallel sides.

T P RH R S   2. All four angles are right angles.

T P RH R S   3. Diagonals are perpendicular.

T P RH R S   4. Only two sides are parallel.

T P RH R S   5. All the angles can be different.

T P RH R S   6. Diagonals are equal.

T P RH R S   7. Opposite angles are congruent.

T P RH R S   8. Opposite angles are supplementary.

(Answers are on page 143.)

BRAIN TICKLERS—THE ANSWERS

Set # 35, page 127

1. 30 degrees
2. 150 degrees
3. 60 degrees
4. 60 degrees
5. 120 degrees
6. 60 degrees
7. 60 degrees

Set # 36, page 132

1. 110 degrees
2. 70 degrees
3. 70 degrees
4. 110 degrees
5. 70 degrees
6. 30 degrees
7. 40 degrees
8. 60 degrees
9. 90 degrees
10. 90 degrees

Set # 37, page 135

1. 80 degrees
2. 40 degrees
3. 50 degrees
4. 100 degrees
5. 90 degrees

Set # 38, page 137

1. 90 degrees

2. 60 degrees

3. 30 degrees

4. 90 degrees

5. 60 degrees

Set # 39, page 140

1. S, E

2. A, C, E

3. E, C

4. S, E

5. C, E

Super Brain Ticklers, page 141

1. P, RH, R, S

2. R, S

3. RH, S

4. T

5. T

6. R, S

7. P, RH, R, S

8. R, S

Chapter 7

Polygons

A polygon is a two-dimensional closed shape made up of line segments. It has a minimum of three sides. A triangle is a polygon; so is a square, a rectangle, a rhombus, a pentagon, and a hexagon. A circle is not a polygon. It is a closed shape but it's not made up of line segments.

A polygon can have three sides, five sides, or one hundred sides. If all sides of a polygon are the same length, then it is an *equilateral polygon*. A rhombus is an equilateral polygon. All four sides of the rhombus are the same length, but the four angles of a rhombus are not the same size.

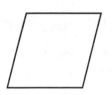

All angles in an *equiangular polygon* have the same measurement. A rectangle is an equiangular polygon. All four angles are 90 degrees. It is not an equilateral polygon because all four sides are not the same length.

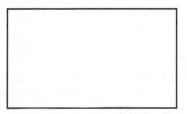

Regular and Irregular Polygons

All the sides and all the angles in a *regular polygon* are equal. A square is a regular polygon.

All parallelograms are polygons. A square, a rhombus, and a rectangle are all polygons. A square is a regular polygon, but a rhombus and a rectangle are irregular.

A polygon is called *irregular* when the sides or angles are not equal.

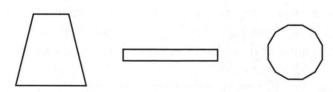

Convex and Concave Polygons

Polygons are also classified as concave or convex. A *convex polygon* is a polygon where all the angles are less than 180 degrees. This trapezoid, a rectangle, and a dodecagon are all convex polygons.

Concave polygons look entirely different. In a concave polygon, at least one of the angles in the polygon is greater than 180 degrees. The arrow on the next page is a concave polygon. There are seven angles in this arrow. Two of the angles are greater than 180 degrees. The arrow is also an irregular polygon.

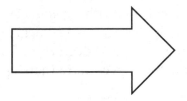

Sides and Angles of Polygons

Polygons are named by the number of sides.

An equiangular triangle is a regular polygon with three sides. Each of the angles of an equilateral triangle is 60 degrees. The total number of degrees in a triangle is 180 degrees.

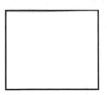

A square is a regular polygon with four equal sides. Each of the interior angles of a square is 90 degrees. The total number of degrees in a square is 4 × 90 degrees, or 360 degrees. A square has 180 degrees more than a triangle.

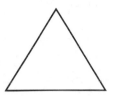

A regular pentagon is a polygon with five equal sides. Each of the interior angles of a regular pentagon is 108 degrees. The total number of degrees in a pentagon is 5 × 108 degrees, or 540 degrees. A pentagon has a total of 180 degrees more than a square.

A regular hexagon is a polygon with six equal sides. Each of the interior angles of a regular hexagon is 120 degrees. The total number of degrees in a hexagon is 6 × 120 degrees, or 720 degrees. A hexagon has a total of 180 degrees more than a pentagon.

A regular heptagon is a polygon with seven equal sides. Each of the angles of a regular heptagon is 128.571 degrees. The total number of degrees in a heptagon is 900 degrees. A heptagon has a total of 180 degrees more than a hexagon.

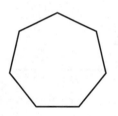

A regular octagon is a polygon with eight equal sides. Each of the angles of a regular octagon is 135 degrees. The total number of degrees in an octagon is 1,080 degrees. An octagon has a total of 180 degrees more than a heptagon.

It is possible to compute the sum of interior angles of any regular polygon. Just subtract 2 from the number of sides in the polygon, and multiply the result by 180. Watch—it's painless.

THEOREM   The sum of the interior angles of a regular polygon = (n – 2)180 degrees, where n is the number of sides in the polygon.

Example:

What is the sum of the angles in a regular hexagon?

Sum = $(n - 2)180$, where n is the number of sides in a hexagon.

Sum = $(6 - 2)180$ degrees

Sum = $4(180) = 720$ degrees

The sum of the angles of a regular hexagon is 720 degrees.

Example:

What is the sum of the angles in a regular dodecagon? Hint: A dodecagon is a polygon with twelve sides.

Sum = $(n - 2)180$, where n is the number of sides in a dodecagon.

Sum = $(12 - 2)180$

$10(180) = 1,800$

The sum of the angles of a regular dodecagon is 1,800 degrees.

Example:

What is the sum of the angles in a regular polygon with twenty sides?

Sum = $(n - 2)180$, where n is the number of sides in the polygon.

Sum = $(20 - 2)180$

$18(180) = 3,240$

The sum of the angles of a regular polygon with 20 sides is 3,240 degrees.

Once you know the total number of degrees inside a regular polygon, it's easy to figure out how many degrees in each angle.

THEOREM   Each interior angle (of a regular polygon) = $\dfrac{180(n - 2)}{n}$ degrees, where n is the number of sides in the polygon.

To compute the size of the interior angles of a regular polygon, follow these two painless steps.

Step 1: Compute the total number of degrees in the regular polygon.

Step 2: Divide the total number of degrees by the number of sides.

Example:

Compute the size of each interior angle of a regular octagon. An octagon has eight sides.

Step 1: Find the total number of degrees in an octagon.

Total degrees $= (n - 2)180$, where n is the number of sides in an octagon.

Total degrees $= (8 - 2)180$

Total degrees $= 6(180) = 1,080$

Step 2: Divide the total number of degrees by the number of sides.

The measure of one interior angle $= \dfrac{1,080}{8}$.

The measure of one interior angle of a regular octagon is 135 degrees.

Example:

Compute the size of each interior angle in a regular polygon with thirty sides.

Step 1: Find the total number of degrees in the polygon.

Total degrees $= (n - 2)180$, where n is the number of sides.

Total degrees $= (30 - 2)180$

Total degrees $= 28(180) = 5,040$

Step 2: Divide the total number of degrees by the number of sides in the polygon.

The measure of one interior angle $= \dfrac{5,040}{30}$.

The measure of one interior angle of a regular polygon with thirty sides is 168 degrees.

⚡ **CAUTION—Major Mistake Territory!**

The size of the interior angles of a polygon can only be computed if the polygon is a regular polygon. The formula will not work for an irregular polygon because all the angles won't be the same size.

Example: There are five interior angles and 540 degrees in every pentagon. To find the number of degrees in the interior angles of a regular pentagon, divide 540 by 5. Every angle of a regular pentagon is 108 degrees. The size of the individual angles of an irregular pentagon are unknown.

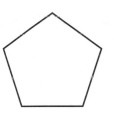

Pentagon A
Irregular Pentagon
Angle Size Unknown

Pentagon B
Regular Pentagon
Angle Size 108 degrees

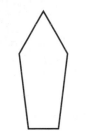 **BRAIN TICKLERS   Set # 40**

1. Find the total number of degrees in a regular polygon with 11 sides.

2. Find the total number of degrees in a regular polygon with 100 sides.

3. Find the number of degrees of a single angle in a regular polygon with 15 sides.

4. Find the number of degrees of a single angle in a regular polygon with 25 sides.

(Answers are on page 157.)

Exterior Angles of Polygons

For each interior angle of a regular polygon there is a corresponding exterior angle. The exterior angle is the angle between any side of the polygon and a line extended from the next side of the polygon.

This square is a polygon. Extending each of the sides of the square in one direction creates four exterior angles, angles *a*, *b*, *c*, and *d*. Each of these angles is a right angle. The sum of these four angles is 360 degrees.

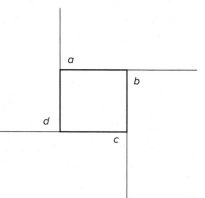

Each of the sides of this hexagon has been extended in one direction to create six exterior angles, angles *a*, *b*, *c*, *d*, *e*, and *f*. Since all the interior angles are equal, all the exterior angles are equal to each other. The sum of the exterior angles of this hexagon is also 360 degrees.

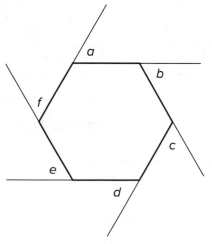

Irregular polygons have exterior angles also. The five sides of this pentagon were extended to create five exterior angles, angles *a*, *b*, *c*, *d*, and *e*. The five interior angles of this pentagon are not equal, so the five exterior angles of this pentagon are not equal.

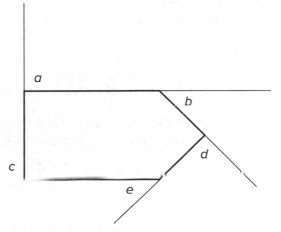

The sum of the exterior angles of a regular polygon will always equal 360 degrees no matter how many sides the polygon has.

THEOREM   To measure an exterior angle of a regular polygon, divide 360 by the number of sides of the polygon.

$$\text{Size of exterior angle of a regular polygon} = \frac{360}{n}$$

(*n* is equal to the number of sides on the polygon)

Example:
What is the size of the exterior angle of a regular hexagon?
Divide 360 by 6, which is the number of sides in the hexagon.

$$\frac{360}{6} = 60$$

Example:
What is the measure of the exterior angle of a regular polygon with ten sides?
Divide 360 by 10, which is the number of sides in the polygon.

$$\frac{360}{10} = 36$$

THEOREM   The sum of an interior angle of a regular polygon and its corresponding exterior angle is always 180 degrees.

If the interior angle of a regular pentagon is 108 degrees, then the corresponding exterior angle is $180 - 108 = 72$. If the exterior angle of a regular octagon is 45 degrees, then the corresponding interior angle is $180 - 45 = 135$.

If you know the interior angle of a regular polygon, it's easy to calculate the number of sides. Just follow these two painless steps.

Step 1:   Calculate the size of the exterior angles of a polygon by subtracting the size of the interior angle from 180.

Step 2:   Divide 360 by the size of a single exterior angle.

Example:

The interior angles of a regular polygon are each 140 degrees. How many sides does the polygon have?

Step 1:   Calculate the size of a single exterior angle by subtracting the size of the interior angle from 180.

$$180 - 140 = 40$$

Step 2:   Divide 360 by the size of a single exterior angle.

$$\frac{360}{40} = 9$$

This polygon has nine sides. It's a nonagon!

Example:

The interior angles of a regular polygon are each 135 degrees. How many sides does the polygon have?

Step 1:   Calculate the size of the exterior angle.

$$180 - 135 = 45$$

Step 2:   Divide 360 by the size of a single interior angle.

$$\frac{360}{45} = 8$$

This polygon has eight sides. It's an octagon!

BRAIN TICKLERS   Set # 41

1. If the interior angle of a regular polygon is 170 degrees, what is the measure of the corresponding exterior angle?

2. If the interior angle of a regular polygon is 30 degrees, what is the measure of the corresponding exterior angle?

3. If the interior angle of a regular polygon is 150 degrees, how many sides does the polygon have?

4. If the interior angle of a regular polygon is 160 degrees, how many sides does the polygon have?

(Answers are on page 157.)

Inscribed and Circumscribed Polygons

The words *inscribe* and *circumscribe* have special meanings in geometry. To inscribe means to put one shape inside another shape. To circumscribe a shape means to put a figure around it. It's possible to inscribe regular polygons inside circles. It's also possible to inscribe circles inside polygons.

This regular pentagon is inscribed in the circle. The vertices of the pentagon are points on the circle. The center of the circle is the center of the pentagon.

This circle is inscribed in a regular hexagon. The sides of the hexagon are tangent to the circle. The center of the circle is the center of the hexagon. This hexagon is circumscribed around this circle. All six sides of the hexagon are tangent to the circle.

This square is both inscribed inside the larger circle and circumscribed around the smaller circle. The center of the square is the center of both of the circles. The sides of the square are tangent to the small circle. The vertices of the square lie on the large circle.

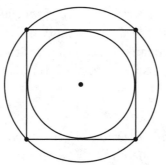

the square

SUPER BRAIN TICKLERS

1. Find the total number of degrees in a regular polygon with fifteen sides.

2. Find the number of degrees in a single angle of a regular polygon with seventeen sides.

3. If the interior angle of a regular polygon is 40 degrees, what is the measure of the corresponding exterior angle?

4. If the interior angle of a regular polygon is 144 degrees, how many sides does the polygon have?

(Answers are on page 157.)

BRAIN TICKLERS—THE ANSWERS

Set # 40, page 151

1. 1,620 degrees

2. 17,640 degrees

3. 156 degrees

4. 165.6 degrees

Set # 41, page 155

1. 10 degrees

2. 150 degrees

3. 12 sides

4. 18 sides

Super Brain Ticklers, page 156

1. 2,340 degrees

2. 158.82 degrees

3. 140 degrees

4. 10 sides

Chapter 8

Circles

Circles are everywhere. A wheel is in the shape of a circle. So are the face of a clock, the rim of a plate, and the lip of a cup. Circles are unique geometric shapes. Every circle has three things in common.

1. Every circle has a center. The center is a fixed point in the middle of the circle.

2. Every point the exact same distance from the center of the circle is on the circle.

3. All the points in a circle are on a single plane.

Circles are usually named by their center. This is circle Q.

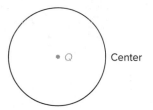

Center

EXPERIMENT

Learn a creative way to make a circle.

Materials
Piece of string 5 inches long
Two pencils
Paper

Procedure

1. Tie one end of the piece of string tightly around a pencil.

2. Tie the other end of the string around a second pencil.

3. Place one of the pencils, point down, firmly in the center of a piece of paper.

4. Keeping the string tight, draw a circle by moving the second pencil around the first.

5. Use a different length of string and draw a different-size circle with the same center.

Something to think about . . .

How does a compass work to draw circles?
What else could you use to draw a circle?

Concentric circles are circles with the same center.

Radius and Diameter

The *radius* of a circle is a line segment that starts at the center of the circle and ends on the circle. Every radius of the same circle is exactly the same length.

Example:

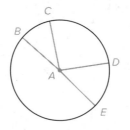

$\overline{AB}, \overline{AC}, \overline{AD}$, and $\overline{AE}$ are all radii of the circle.
All these radii are the same length.
$\overline{AB} = \overline{AC} = \overline{AD} = \overline{AE}$

The *diameter* of a circle is a line segment that connects both sides of the circle and passes through the center of the circle. All diameters of a single circle are the same length.

Example:

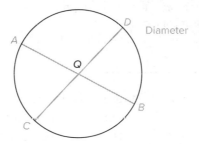

Diameter

$\overline{AB}$ and $\overline{CD}$ are both diameters of circle Q.
$\overline{AB} = \overline{CD}$.

EXPERIMENT

Discover the relationship between the radius and the diameter of a circle.

Materials
Compass
Paper
Pencil
Ruler

Procedure
1. Use a compass to draw a large circle.

2. Use the compass to draw a smaller circle.

3. Draw three lines through the center of each circle. These are the diameters of each circle.

4. Measure the length of each diameter. Enter the results in the chart.

5. Draw three radii in each circle.

6. Measure the length of each radius. Enter the results in the chart.

Radius/diameter	Small circle	Large circle
Radius 1		
Radius 2		
Radius 3		
Diameter 1		
Diameter 2		
Diameter 3		

> **Something to think about . . .**
> Are all the radii from one circle the same length?
> Are all the diameters from one circle the same length?
> What is the relationship between the length of the radius and the length of the diameter?

In geometry, "if and only if" statements are common. "If and only if" statements mean the first statement will always be true when the second is true, and will only be true if the second is true.

THEOREM   Two circles are congruent if and only if their radii are the same length.

This theorem means two things:

1. Two circles are congruent only if their radii are the same length. In other words, the circles cannot be congruent if the radii are not the same length.

2. Two circles are congruent if their radii are the same length. Proving the radii are congruent is sufficient to prove that the circles are congruent.

You can prove that two circles are congruent if you can prove that either their radii or diameters are congruent.

BRAIN TICKLERS   Set # 42

1. If a circle has a 3-inch radius, what is the length of the diameter?

2. If a circle has a 12-inch diameter, what is the length of the radius?

(Answers are on page 191.)

Regions of a Circle

A circle divides a plane into three distinct areas:

- interior of a circle
- exterior of a circle
- the circle itself

The area inside the circle is called the *interior of the circle*. Points A, B, and C lie in the interior of the circle.

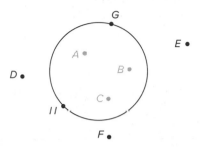

Points G and H lie on the circle. The *exterior of the circle* is all the points that are farther from the center of the circle than the length of the radius of the circle. Points D, E, and F lie in the exterior of a circle.

The distance from the center of the circle to any point inside the circle is less than the length of the radius. The distance from the center of the circle to any point in the exterior of the circle is greater than the length of the radius. The distance from the center of the circle to any point on the circle is exactly the length of the radius.

BRAIN TICKLERS   Set # 43

1. A circle has a radius of 4 inches. A point lies 6 inches from the center of the circle. Is this point on the circle, in the interior of the circle, or in the exterior of the circle?

2. A circle has a diameter of 10 inches. A point lies 8 inches from the center of the circle. Is this point on the circle, in the interior of the circle, or in the exterior of the circle?

3. A circle has a diameter of 12 inches. A point lies 6 inches from the center of the circle. Is this point in the interior of the circle, in the exterior of the circle, or on the circle?

4. A circle has a radius of 5 inches. A point lies 3 inches from the center of the circle. Is this point in the interior of the circle, in the exterior of the circle, or on the circle?

(Answers are on page 191.)

Pi

π is a Greek symbol and is spelled as "pi" and pronounced "pie." π is determined by dividing the circumference of a circle by its diameter. π is 3.141592 . . . and continues forever. π is a nonterminating, nonrepeating decimal. When you solve problems where you need the number π, use an approximation of π. The number 3.14 is a common approximation of π. So is $\frac{22}{7}$. When π is involved, compute two decimal places.

EXPERIMENT

Determine the relationship between the circumference and the diameter of a circle.

Materials
String
Ruler
Pencil
Paper
Calculator
Plate, saucer, cup, glass, garbage pail, quarter, and other round items

Procedure

1. Measure the distance around each round object by putting a string around each object. Measure the length of the string used to circle each item using the ruler. Write your findings in the chart.

2. Measure the length of the diameter using the ruler and the string. Make sure that the diameter goes through the center of the circle. Write your findings in the chart.

3. Divide the distance around each circle (circumference) by the distance across the center of the circle (diameter). Write your results in the chart.

4. Compare the results of column 3 for the glass, cup, saucer, quarter, plate, and garbage pail.

Round item	Length of circumference	Length of diameter	Length of circumference divided by length of diameter
Glass			
Cup			
Saucer			
Quarter			
Plate			
Round garbage pail			

Something to think about . . .
If the diameter of one circle is twice as large as another circle, what is the relationship between the circumferences of these two circles?

Circumference

The distance around a circle is called the circumference and is measured in linear units.

To figure out the circumference of a circle, follow these painless steps:

Step 1: Find the diameter of the circle.

Step 2: Multiply the diameter by π.

Circumference of a Circle

$$C = \pi d$$

d stands for the diameter of a circle.

To approximate the circumference of a circle,
use either of these formulas:

$$C = d\left(\frac{22}{7}\right) \quad \text{or} \quad C = d(3.14)$$

Example:

Compute the circumference of a circle with diameter 4.

$$C = 3.14d = 3.14(4) = 12.56 \text{ or}$$
$$C = \left(\frac{22}{7}\right)(d) = \left(\frac{22}{7}\right)4 = \frac{8}{7} = 12\frac{4}{7}$$

Computing the circumference of a circle using the radius is painless.

Step 1: Find the radius of the circle.

Step 2: Multiply the radius by 2 since the diameter is twice the radius.

Step 3: Multiply the diameter by π.

Circumference of a Circle
$$C = 2\pi r$$

Example:

Compute the circumference of a circle with radius 1.

Step 1: Find the radius of the circle.

The radius is 1.

Step 2: Multiply the radius by 2 to find the diameter of the circle.

$(2)(1) = 2$

The diameter is 2.

Step 3: Multiply the diameter by π.

$$2\left(\frac{22}{7}\right) = \frac{44}{7} = 6\frac{2}{7} \text{ or}$$

$2(3.14) = 6.28$

The circumference of the circle is 2π or $6\frac{2}{7}$ or 6.28.

THEOREM   Congruent circles have the same circumference.

BRAIN TICKLERS   Set # 44

1. Find the circumference of a circle with diameter 10 in terms of π.

2. Find the circumference of a circle with radius 5 in terms of π.

3. Find the circumference of a circle with radius $\frac{1}{2}$ in terms of π.

4. Find the radius of a circle with circumference 6π.

(Answers are on page 191.)

Area of a Circle

The *area* of a circle is the area covered by the interior of the circle and the circle itself.

EXPERIMENT

Learn to approximate the area of a circle.

Materials
Graph paper
Pencil
Compass

Procedure
1. Draw a circle on a piece of graph paper.

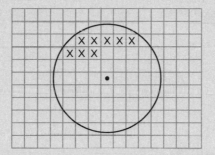

2. Count the number of complete squares inside the circle. Put an x on each square you count to make sure that you don't count the same square twice. Write this number down.

3. Count the number of partial squares inside the circle. Shade each of these partial squares. Multiply the number of partial squares by $\frac{1}{2}$. Write this number down.

4. Add the numbers in Step 2 and Step 3.

Something to think about . . .
Is there an easier way to find the area of a circle?

Finding the area of a circle is painless, especially if you state the answers in terms of π.

Step 1: Find the radius of the circle.

Step 2: Multiply π times the radius squared.

Area of a Circle
$$A = \pi r^2$$

Example:

Find the area of a circle with radius 3.

Area of a circle $= \pi(3)^2$
Area of a circle $= 9\pi$ square units

You can also find the area of a circle if you know the diameter. Just square half the diameter and multiply the result by π.

Area of a Circle
$$A = \pi\left(\frac{d^2}{2}\right)$$

Example:

What is the area of a circle with a diameter of 8 inches?

First find the radius of the circle by multiplying the diameter by $\frac{1}{2}$.

$\frac{1}{2}(8) = 4$ inches

Area of a circle $= \pi(4)^2$
Area of a circle $= 16\pi$ square inches

You can also find the area of a circle if you know the circumference. Just follow these painless steps:

Step 1: First divide the circumference by π to find the diameter.

Step 2: Divide the diameter by 2 to find the radius.

Step 3: Square the radius of the circle.

Step 4: Multiply the squared radius by π.

Example:

Find the area of a circle with circumference 10π.

Step 1: First divide the circumference by π to find the diameter.
10π divided by π is 10.
The diameter of the circle is 10.

Step 2: Divide the diameter by 2 to find the radius of the circle.

10 divided by 2 is 5.
The radius of the circle is 5.

Step 3: Square the radius of the circle.
$$5^2 = 25$$

Step 4:   Multiply the squared radius by π.

$$25\pi$$

The area of the circle is 25π square units.

BRAIN TICKLERS   Set # 45

1. What is the area of a circle with radius 10?

2. What is the area of a circle with diameter 2?

3. What is the area of a circle with circumference 20π?

(Answers are on page 191.)

EXPERIMENT

Compare the area of a circle with the area of a square.

Materials
   Graph paper
   Pencil
   Calculator

Procedure
   1. Draw a square with a side of 10 units on the graph paper.

   2. Inscribe a circle inside the square. To inscribe a circle in a square, each side of the square should be touched by the circle once. This circle has a diameter of 10 units.

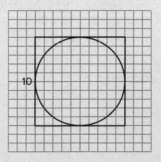

   3. Compute the area of the square. (Area $= s \times s$)

   4. Computer the area of the circle. (Area $= 3.14 \times r \times r$)

   5. Which area is larger?

6. Divide the area of the circle by the area of the square. Multiply the answer by 100 to change the answer to a percent.

7. What percent of the area of the square is the area of the circle?

8. Repeat the first seven steps for a circle with a diameter of 2 units and a square with side 2.

Something to think about . . .

How can the results be used to estimate the area of a circle?

Inscribe a square in a circle. What percentage of the area of a circle is the area of a square?

THEOREM   Congruent circles have the same area. If two circles have the same area, they must be congruent.

To prove two circles congruent, show that their

- radii are the same length.
- diameters are the same length.
- circumferences are the same length.
- areas are the same.

Degrees in a Circle

A right angle has a measure of 90 degrees. The measure of a straight angle is 180 degrees. One circular rotation equals 360 degrees.

The degrees of a circle can be proven by measuring the central angles of a circle. A *central angle* is an angle whose vertex is the center of the circle and whose sides intersect the sides of a circle.

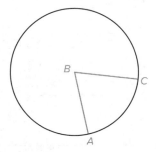

Angle *ABC* is a central angle. Use a protractor to measure a central angle.

EXPERIMENT

Discover the number of degrees in a circle.

Materials
Paper
Pencil
Protractor

Procedure
1. Measure each of the central angles in these circles using a protractor. Enter the results in the chart.

2. Add the measures of the angles. Enter your answer in the chart.

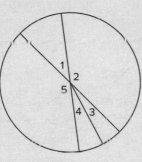

Circle 1

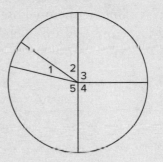

Circle 2

Angle	Circle 1	Circle 2
Angle 1		
Angle 2		
Angle 3		
Angle 4		
Angle 5		
Total of angles 1, 2, 3, 4, and 5		

Something to think about . . .

Draw a small circle approximately half the size of Circles 1 and 2. How many degrees are in this circle?

Draw a large circle approximately twice the size of Circles 1 and 2. How many degrees are in this circle?

THEOREM   Every circle has a total of 360 degrees.

BRAIN TICKLERS   Set # 46

1. How many degrees are in a circle with radius 4?

2. How many degrees are in a circle with diameter 10?

3. How many degrees are in a circle with radius 100?

(Answers are on page 191.)

Chords, Tangents, and Secants

A *chord* is a line segment whose endpoints lie on the circle. Segments *AB*, *CD*, *EF*, and *GH* are all chords.

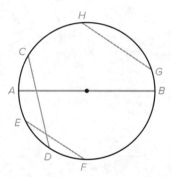

Notice that segment *AB* is both a chord and a diameter of the circle.

Which of these segments are chords?

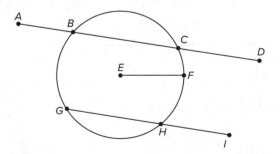

Segments *AD* and *GI* extend beyond the circle and are not chords. One of these segments (*EF*) is the radius of the circle. It contacts only one side of the circle. However, segments *BC* and *GH* are chords.

A *tangent* is a line on the exterior of the circle that touches the circle at exactly one point.

Example:

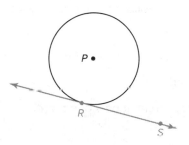

Line *RS* is tangent to circle *P*.
The circle and the line intersect at exactly one point, *R*

A *secant* is a line that intersects a circle at two points. Lines *AB* and *CD* are secants. These secants are parallel lines and do not intersect.

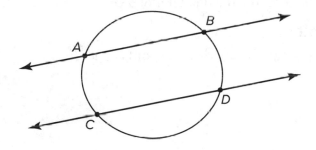

Lines *JK* and *LM* are also secants. They intersect at point *N*.

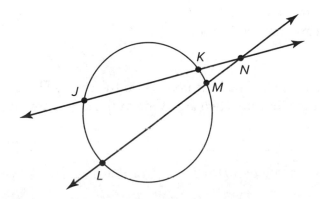

BRAIN TICKLERS   Set # 47

Decide whether each of the following statements is true or false.

1. A radius is a chord.

2. A tangent intersects a circle at two points.

3. A diameter is a chord.

4. A secant intersects a circle at only one point.

(Answers are on page 192.)

Arcs

An *arc* is a portion of a circle.

A *semicircle* is an arc that is exactly half a circle. The endpoints of a semicircle lie on the same diameter of a circle.

Example:
Draw a circle. Draw a diameter and call it $\overline{AB}$.

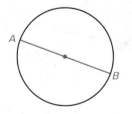

Each semicircle is called $\overset{\frown}{AB}$.
Not all arcs are semicircles.

Example:
Draw a circle and pick two points on the circle. These two points divide the circle into two arcs. One of the arcs is smaller than the other arc.

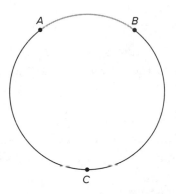

The smaller arc is called the *minor arc*. A minor arc is less than a semicircle. Label the minor arc using two letters. Place an arc sign over the letters. Arc *AB* is written as $\overset{\frown}{AB}$.

The larger arc is called a *major arc*. A major arc is larger than a semicircle. Label the major arc using three letters by picking another letter on the arc. Three letters are used to label major arcs so they are not confused with the minor arcs. Place an arc sign over the three letters. Arc *ACB* is written as $\overset{\frown}{ACB}$.

BRAIN TICKLERS   Set # 48

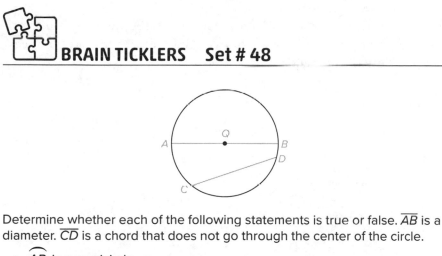

Determine whether each of the following statements is true or false. $\overline{AB}$ is a diameter. $\overline{CD}$ is a chord that does not go through the center of the circle.

1. $\overset{\frown}{AB}$ is a semicircle.

2. $\overset{\frown}{CD}$ is a major arc.

3. $\overset{\frown}{CAB}$ is greater than a semicircle.

(Answers are on page 192.)

Measuring Arcs

The measure of an arc is expressed in degrees, while the length of an arc is expressed in linear measurements such as inches, centimeters, and feet.

Finding the measure of an arc is easy. Just measure the central angle that the arc cuts off. Remember that a central angle is an angle that has its vertex as the center of the circle and its sides are radii of the circle.

To find the measure of an arc, just follow these painless steps:

Step 1: Draw a line segment from one end of the arc to the center of the circle.

Step 2: Draw a line segment from the other end of the arc to the center of the circle.

Step 3: Measure the angle created by the two line segments. If the arc is a minor arc (less than 180 degrees), this is the measure of the arc.

Step 4: If the arc is a major arc (greater than 180 degrees), subtract the measure of the angle from 360 degrees to find the measure of the major arc.

Example:

Find the measure of the arc indicated.

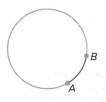

Step 1:   Connect one end of the arc to the center of the circle.

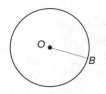

Step 2: Connect the other end of the arc to the center of the circle.

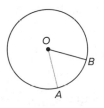

Step 3: Measure the angle created by the two line segments.

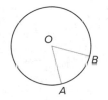

Angle *AOB* is 60 degrees.

Step 4: If the arc is a major arc (greater than 180 degrees), subtract the measure of the angle from 360 degrees to find the measure of the major arc. Since the arc is a major arc, subtract 60 degrees from 360 degrees.

$$360 - 60 = 300$$

The measure of the arc is 300 degrees.

To find the length of an arc, just follow these painless steps:

Step 1: Find the circumference of the circle.

Step 2: Divide the number of degrees in the arc by 360.

Step 3: Multiply the circumference of the circle by the ratio found in Step 2.

Example:

Find the length of an arc with a measure of 30 degrees that is part of a circle with radius 12 inches.

Step 1:   Find the circumference of the circle.

The circumference of the circle $= \pi$ (diameter).
If the radius is 12, the diameter is 24.
The circumference is 24π.

Step 2:   Divide the number of degrees in the arc by 360 degrees.

The arc is 30 degrees. So, 30 degrees divided by 360 degrees is $\frac{1}{12}$.

Step 3:   Multiply the circumference by the ratio.
Multiply $24\pi\left(\frac{1}{12}\right)$.

The answer is 2π or 6.28.

A *semicircle* is half a circle. The endpoints of a semicircle are a diameter of a circle. A semicircle has an arc of 180 degrees.

To find the circumference of a semicircle, follow these four painless steps:

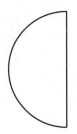

Step 1: Find the circumference of the entire circle.

Step 2: Divide the circumference by 2 to find the length of the arc of the semicircle.

Step 3: Find the length of the diameter.

Step 4: Add the length of the arc to the length of the diameter.

Example:

Find the circumference of a semicircle with diameter 10.

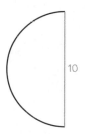

Step 1: Find the circumference of the entire circle. The circumference of the entire circle is 10(3.14) or 31.4.

Step 2: Divide the circumference by 2 to find the length of the arc of the semicircle.

$$\frac{31.4}{2} = 15.7$$

Step 3: Find the length of the diameter.
The length of the diameter is 10.

Step 4: Add the length of the arc (15.7) to the length of the diameter (10) to find the circumference of the semicircle.

$$15.7 + 10 = 25.7$$

THEOREM Congruent arcs have the same degree measure and the same length.

$\overarc{AB}$ and $\overarc{CD}$ have both the same measure and the same length.

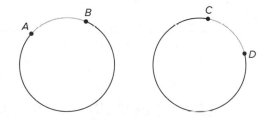

$\overarc{EF}$ and $\overarc{GH}$ are not congruent. They have the same length but not the same measure.

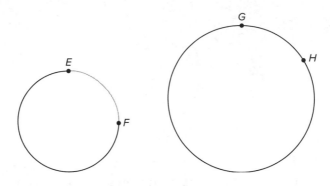

$\overset{\frown}{IJ}$ and $\overset{\frown}{KL}$ are not congruent. They have the same measure but not the same length.

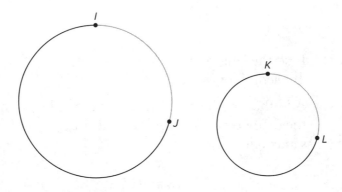

CAUTION—Major Mistake Territory!

Congruent arcs must come from congruent circles. Just because two arcs contain the same number of degrees does not mean that they are congruent.

BRAIN TICKLERS Set # 49

1. Find the length of an arc that is 180 degrees in a circle with diameter 4.

2. Find the length of an arc that is 270 degrees in a circle with diameter 10.

3. Find the length of an arc that is 45 degrees in a circle with diameter 2.

(Answers are on page 192.)

Angles and Circles

Eight different types of angles are formed when secants, tangents, radii, or chords of a circle intersect. The measure of each of these angles depends on the measure of the intersected arc. The vertex of these angles can be located inside a circle or outside a circle, or it can intersect a circle.

Case I: Two radii

An angle formed by two radii of a circle is called a central angle. The vertex of a central angle is the center of the circle. The measure of the central angle is equal to the measure of the intersected arc. Any two radii inside a circle will form a central angle.

Example:

The measure of angle BAC = measure of arc BC.

If arc BC = 45 degrees then angle BAC = 45 degrees.

Likewise if angle BAC = 45 degrees, then arc BC = 45 degrees.

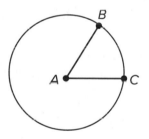

Case II: Two chords

An angle formed by two chords is called an inscribed angle. The vertex of the inscribed angle is on the circle. The measure of an inscribed angle is equal to half the measure of the intersected arc.

Example:

Angle ABC is formed by two chords.

The measure of angle $ABC = \frac{1}{2}$ arc AC.

If arc AC = 80 degrees, then angle $ABC = \frac{1}{2}(80) = 40$.

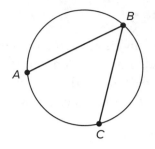

Case III: Two tangents

The vertex on an angle formed by the intersection of two tangents is always outside the circle. The measure of an angle formed by two tangents is half the difference of the two intersected arcs. A third point on the circle (point D) is placed to differentiate between the major arc and the minor arc.

To find the measure of angle BAC, follow these three painless steps.

Step 1: Find the measure of both arcs.

Step 2: Subtract the measure of the minor arc from the measure of the major arc.

Step 3: Take half the difference.

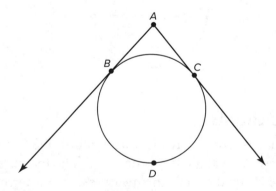

Example:

Find the measure of angle BAC if arc $BC = 90$ degrees.

Step 1: Find the measure of arc BC and arc BDC.

If arc BC is 90 degrees, arc BDC is $360 - 90 = 270$ degrees.

Step 2: Subtract the measure of arc BC from the measure of arc BDC.

$270 - 90 = 180$

Step 3: Take half the difference.

$\frac{1}{2}(180) = 90$

Angle $BAC = 90$ degrees.

Case IV: One tangent and one secant

A tangent and a secant can intersect outside the circle or on the circle. Since tangents have no points inside the circle, they never intersect inside the circle. This tangent and secant intersect outside the circle. The tangent intersects the circle at exactly one point, C. The secant intersects the circle at two points, B and D. Point A, where the secant and the tangent intersect, is the vertex of the angle. The measure of an angle formed by a tangent and a secant that intersect outside the circle is half the difference of the two intersected arcs.

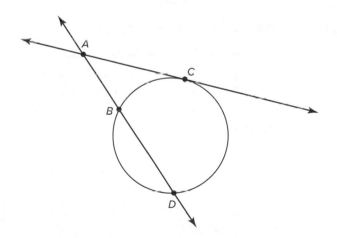

To find the measure of an angle formed by a tangent and a secant that intersect outside the circle, follow these three steps.

Step 1: Find the measure of both arcs.

Step 2: Subtract the measure of the smaller arc from the measure of the larger arc.

Step 3: Take half the difference.

   Example:

   Find the measure of angle *BAC* if the measure of arc
   BC = 80 degrees, and the measure of arc *BD* = 100 degrees.

   Measure of angle *BAC* = $\frac{1}{2}$(measure of arc *CD* − measure of
   arc *BC*)

Step 1: Find the measure of both arcs. Arc *BC* = 80. Since
   there are 360 degrees in a circle, arc *CD* = 360 −
   (100 + 80) = 180.

Step 2: Subtract the measure of arc *BC* from the measure of
   arc *CD*.

   180 − 80 = 100

Step 3: Take half the difference.

   Angle *A* = $\frac{1}{2}$(100)

   Angle *A* = 50 degrees.

Case V: One tangent and one secant

A tangent and a secant can form an angle by intersecting outside the
circle, but they can also intersect the circle at the point of tangency.
If a tangent and a secant intersect at the point of tangency, two
angles are formed. The measure of each angle is equal to half the
measure of the intersected arc.

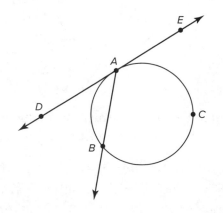

Tangent DE and secant BA intersect at point A. Two angles are formed: angle BAD and angle BAE.

Angle $BAD = \dfrac{1}{2}$ the measure of arc AB

Angle $BAE = \dfrac{1}{2}$ the measure of arc BCA

To find the measure of each of the two angles, follow these two steps.

Step 1: Find the measure of both of the intersected arcs.

Step 2: Take half of the measure of each of the intersected arcs.

Example:

Find the measure of angle BAD and the measure of angle BAE if the measure of arc $AB = 100$ degrees.

Step 1: Find the measure of arc AB and arc ACB.

Arc $AB = 100$

Arc $ACB = 360 - 100 = 260$

Step 2: Take half the measure of each of the intersected arcs.

Angle $BAD = \dfrac{1}{2}(100) = 50$ degrees

Angle $BAE = \dfrac{1}{2}(260) = 130$ degrees

Case VI: Two secants

Two secants can intersect outside of the circle, inside the circle, or even on the circle. The measure of an angle formed by two secants that intersect outside the circle is half the difference of the measure of the two intersected arcs.

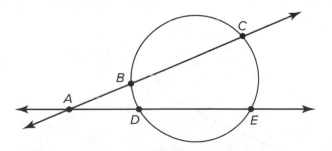

Secants AC and AE intersect each other outside the circle.

The measure of angle $CAE = \frac{1}{2}$(the measure of arc CE − the measure of arc BD).

To find the measure of angle BAD, just follow these three painless steps.

Step 1: Find the measure of both arcs.

Step 2: Subtract the measure of the smaller arc from the measure of the larger arc.

Step 3: Take half the difference.

> ### Example:
> Find the measure of angle CAE if arc $BD = 30$ degrees and arc $CE = 70$ degrees.
>
> **Step 1:** Find the measure of both arcs.
>   Arc $BD = 30$ and arc $CE = 70$
>
> **Step 2:** Subtract the measure of the smaller arc from the measure of the larger arc.
>   Arc CE − arc $BD = 70 − 30 = 40$
>
> **Step 3:** Take half the difference.
>   Angle $CAE = \frac{1}{2}(40) = 20$
>   Angle $CAE = 20$ degrees

Case VII: Two secants

Two secants can also intersect in the interior of a circle. When they do, they form two pairs of vertical angles. The measure of each angle is half the sum of the intersected arcs of the angle and its vertical angle.

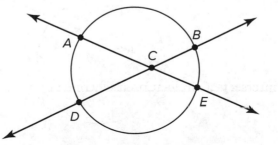

To find the measure of each of the four angles, follow these three painless steps.

Step 1: Find the measure of the arcs of a pair of vertical angles.

Step 2: Take half the sum of the two arcs to find the measure of each of the vertical angles.

Step 3: Subtract the measure of one of the angles from 180 to find the measure of each of the angles in the other pair of vertical angles.

Example:

The secants AE and DB intersect each other inside the circle at point C, forming angles ACD, ACB, BCE, and DCE. Find the measures of angles ACD, BCE, ACB, and DCE, if the measure of arc $AD = 60$ degrees and arc $BE = 40$ degrees.

Step 1: Find the measure of the arcs of two vertical angles.
Arc $AD = 60$ and arc $BE = 40$

Step 2: Take half the sum of the two arcs to find the measure of each of the vertical angles.

Angle $ACD = \frac{1}{2}(60 + 40)$

Angle $ACD = \frac{1}{2}(100)$

Angle $ACD = 50$

Angle $BCE =$ Angle ACD since they are vertical angles

Angle $BCE = 50$ degrees

Step 3: Subtract the measure of one of the angles from 180 to find the measure of each of the angles of the other pair of vertical angles.
Angle $ACB = 180 - 50 = 130$
Angle $DCE =$ Angle $ACB = 130$ degrees.

Case VIII: A tangent and a radius of a circle

A tangent intersects a circle at precisely one point. Only one radius intersects the tangent at that point, forming two right angles. The vertex of each of these angles lies on the circle.

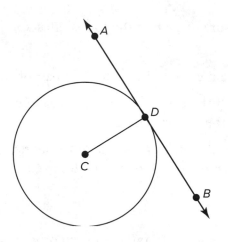

Example:

Tangent *AB* intersects radius *CD* at point *D*. *AB* is perpendicular to *CD*.

Angle *ADC* = 90 degrees and angle *BDC* = 90 degrees

Summary

There are eight possible cases of angles associated with circles.

Angle Formed by	Location	Angle Measure Equals
Two radii	Interior of the circle	= the measure of the intersected arc
Two chords	On the circle	= half the measure of the intersected arc
Two tangents	Exterior of the circle	= half the difference of the two intersected arcs
One tangent and one secant	Exterior of the circle	= half the difference of the intersected arcs
One tangent and one secant	On the circle	= the measure of the intersected arc
Two secants	Exterior of the circle	= half the difference of the two intersected arcs
Two secants	Interior of the circle	= half the sum of the intersected arcs of the angle and its vertical angle
A tangent and a radius	On the circle	= 90 degrees

BRAIN TICKLERS   Set # 50

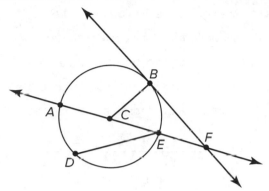

Arc *BE* = 60 degrees

Arc *AD* = 60 degrees

C is the center of the circle.

1. What is the measure of angle *CBF*?

2. What is the measure of angle *BFA*?

3. What is the measure of angle *BCE*?

4. What is the measure of angle *BCA*?

5. What is the measure of angle *AED*?

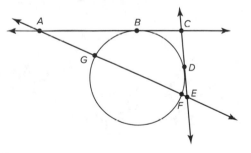

Arc *BD* = 90 degrees

Arc *BG* = 70 degrees

Arc *DF* = 50 degrees

6. What is the measure of angle *ACE*?

7. What is the measure of angle *CEA*?

8. What is the measure of angle *CAE*?

(Answers are on page 192.)

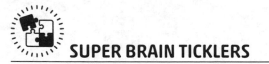

SUPER BRAIN TICKLERS

1. What is the area of a circle with diameter 12?

2. What is the circumference of a circle with radius 4?

3. What is the circumference of a circle with area 9π?

4. What is the area of a circle with circumference 10π?

5. What is the length of a 60-degree arc on a circle with radius 12?

(Answers are on page 193.)

BRAIN TICKLERS—THE ANSWERS

Set # 42, page 162

1. 6 inches

2. 6 inches

Set # 43, page 163

1. Exterior of the circle

2. Exterior of the circle

3. On the circle

4. Interior of the circle

Set # 44, page 166

1. 10π

2. 10π

3. π

4. 3

Set # 45, page 169

1. 100π

2. π

3. 100π

Set # 46, page 172

1. 360 degrees

2. 360 degrees

3. 360 degrees

Set # 47, page 174

1. False
2. False
3. True
4. False

Set # 48, page 175

1. True
2. False
3. True

Set # 49, page 180

1. 2π
2. 7.5π
3. 0.25π

Set # 50, page 189

1. 90 degrees
2. 30 degrees
3. 60 degrees
4. 120 degrees
5. 30 degrees
6. 90 degrees
7. 55 degrees
8. 70 degrees

Super Brain Ticklers, page 190

1. 36π

2. 8π

3. 6π

4. 25π

5. 4π

Chapter 9

Perimeter, Area, and Volume

It's a Matter of Units

It's possible to measure the perimeter, area, surface area, and volume of various shapes. The *perimeter* is the distance around a shape. The perimeter is measured in inches, feet, yards, miles, centimeters, meters, and kilometers.

The *area* is the amount of flat space a flat shape encloses. Square units are used to measure area. Examples of square units are square inches, square feet, square yards, square miles, square centimeters, square meters, and square kilometers. Write square units by putting a small two over the units to indicate "square."

Square inches = in^2
Square feet = ft^2
Square miles = mi^2
Square meters = m^2

Surface area is the outside surface of a solid shape. Surface area is measured in square units.

Volume is the space inside a three-dimensional shape. A liter of water and a cup of sugar are examples of volume. Volume is measured in cubic units. Think of each cubic unit as a little block or cube. Write cubic units by putting a small three over the units.

Cubic inches = in^3
Cubic feet = ft^3
Cubic miles = mi^3

Cubic centimeters = cm^3
Cubic meters = m^3
Cubic kilometers = km^3

It's not hard to learn how to find the perimeter, area, surface area, and volume of common figures.

Triangles

A triangle has three sides and three angles.

Perimeter

The perimeter of a triangle is the sum of the sides of a triangle.

Perimeter of a Triangle = Side 1 + Side 2 + Side 3

Example:

Find the perimeter of a triangle with sides 8, 10, and 12 feet.

Just add the three sides together.
8 + 10 + 12 = 30 feet

Area

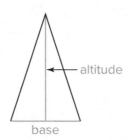

The area of a triangle is the space inside the triangle. The area of a triangle is $\frac{1}{2}$ base × altitude. The *altitude* is the distance from a vertex of the triangle perpendicular to the opposite side. The altitude is also called the *height*.

Area of a Triangle = $\frac{1}{2}$ (base × height)

Example:

What is the area of a triangle with base 8 and altitude 10?

The area is $\frac{1}{2}(8)(10)$.

The area is 40.

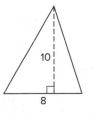

DRAIN TICKLERS   Set # 51

1. Find the perimeter of an equilateral triangle with sides of 5 inches each.

2. Find the perimeter of an isosceles triangle with two sides of 4 inches each and a third side 1 inch long.

3. What's the area of a triangle with base 4 inches and height 1 inch?

(Answers are on page 231.)

Rectangles

A *rectangle* is a parallelogram. It has four right angles and two pairs of parallel sides.

Perimeter

To find the perimeter of a rectangle, just add the length of all four sides together.

Perimeter of a Rectangle $= l + l + w + w$

Example:

Find the perimeter of a rectangle with sides 5 inches and 7 inches.

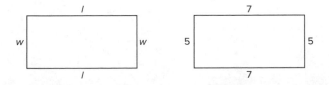

Just add all four sides of the rectangle together.
$7 + 7 + 5 + 5 = 24$
The perimeter is 24 inches.

EXPERIMENT

Learn how to determine the area of a rectangle by counting squares.

Materials
Graph paper
Pencil

Procedure

1. Draw the rectangles indicated on the chart on a piece of graph paper.

2. Count the number of small squares inside each drawn rectangle. Enter the results in the chart.

3. Compute the area of each rectangle by multiplying the length by the width of each rectangle. Enter the results in the chart.

4. Compare the areas you found in Steps 2 and 3.

Rectangle	Length	Width	Area found by counting small squares	Area found by multiplying $l \times w$
Rectangle 1	6	5		
Rectangle 2	10	3		
Rectangle 3	2	4		
Rectangle 4	4	2		
Rectangle 5	1	1		

Something to think about . . .
How could you determine the area of a rectangle with sides 10 inches $\times$ 12 inches?

Area

To determine the area of a rectangle, just multiply the length times the width.

Area of a Rectangle $= lw$

Example:

To find the area of a rectangle 2 inches by 4 inches, just multiply 2 × 4.

The area of this rectangle is 8 square inches.

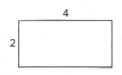

**BRAIN TICKLERS   Set # 52**

1. What is the perimeter of a rectangle with length 10 inches and height 6 inches?

2. What is the perimeter of a rectangle with sides 4 meters and 16 meters?

3. What is the area of a rectangle with length 10 inches and height 5 inches?

4. What is the area of a rectangle with length 4 centimeters and height 2 centimeters?

(Answers are on page 231.)

Squares

A *square* has four right angles and four equal sides. A square is a special type of rectangle. Since all four sides of a square are equal, the length and width of the square are called the sides of the square.

Perimeter

To find the perimeter of a square, just add all the sides together.

Perimeter of a Square = $s + s + s + s = 4s$

Example:

What is the perimeter of a square with sides 9 inches long?

Add all four sides of the square together.
9 + 9 + 9 + 9 = 36 inches

EXPERIMENT

Discover the formula for finding the perimeter of a square.

Materials
Pencil
Paper

Procedure

1. Draw each of the squares listed in the chart.

2. Label each of the sides.

3. Add the sides of each square together to find its perimeter. Enter the results in the chart.

4. Multiply the length of one side of each square by the number of sides (4). Enter the results in the chart.

Square	Length of one side, *s*	Perimeter found by adding the sides	Perimeter found by multiplying 4 times *s*
Square 1	5		
Square 2	7		
Square 3	1		
Square 4	10		

5. Did you get the same result each way? Which way was the easiest?

Something to think about . . .

Can you find a shortcut method to find the perimeter of a rectangle?

Perimeter of a Square $= s + s + s + s$

Perimeter of a Square $= 4(s)$, which is 4 times s

Area

The area of a square is length times width. Since the length and width of a square are exactly the same, the area of a square is $s \times s$ or s^2.

Area of a Square $= s \times s = s^2$

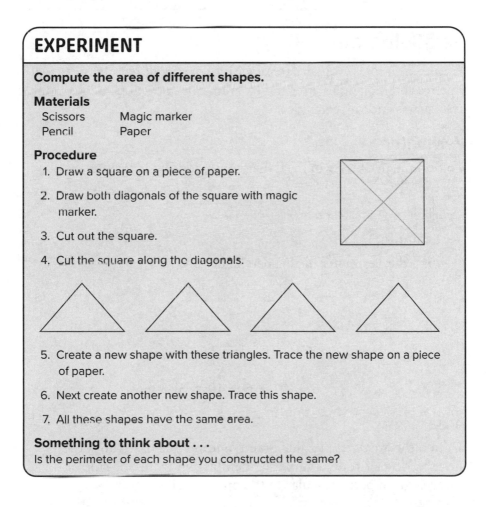

EXPERIMENT

Compute the area of different shapes.

Materials

Scissors   Magic marker
Pencil   Paper

Procedure

1. Draw a square on a piece of paper.

2. Draw both diagonals of the square with magic marker.

3. Cut out the square.

4. Cut the square along the diagonals.

5. Create a new shape with these triangles. Trace the new shape on a piece of paper.

6. Next create another new shape. Trace this shape.

7. All these shapes have the same area.

Something to think about . . .

Is the perimeter of each shape you constructed the same?

BRAIN TICKLERS Set # 53

1. What is the perimeter of a square with side 7?

2. What is the area of a square with side 5?

3. What is the area of a square with perimeter 12?

4. What is the perimeter of a square with area 100 square units?

(Answers are on page 231.)

Parallelograms

A *parallelogram* is a quadrilateral with two pairs of parallel sides. All squares and rectangles are parallelograms, but not all parallelograms are squares and/or rectangles.

Perimeter

To find the perimeter of a parallelogram, add the length of all four sides.

Perimeter of a Parallelogram $= l + w + l + w$

Example:

Find the perimeter of this parallelogram.

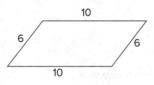

$$6 + 10 + 6 + 10 = 32$$

Area

To find the area of a parallelogram, multiply the base times the altitude. The altitude is a line segment drawn from one side of the parallelogram perpendicular to the opposite side.

Area of a Parallelogram = $l(a)$

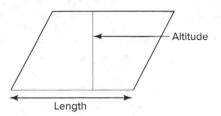

Altitude

Length

Example:

What is the area of a parallelogram with base 8, width 5, and altitude 4?

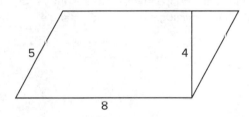

Multiply the base (8) by the altitude (4).
$8 \times 4 = 32$
The area is 32 square units.

⚡ CAUTION—Major Mistake Territory!

Don't compute the area of a parallelogram by multiplying the length times the width. You must multiply the altitude times the length.

EXPERIMENT

Change a parallelogram to a rectangle.

Materials
 Pencil
 Graph paper
 Scissors
 Tape

Procedure
 1. Draw a parallelogram on a piece of graph paper. The parallelogram you draw should *not* be a rectangle.

2. Make the parallelogram 7 squares long and 4 squares tall.

3. Cut out the parallelogram.

4. Cut a triangle off one end of the parallelogram. Start at the inside corner and cut straight down. Look at the diagram.

5. Slide the triangle to the right and tape it to the other end of the parallelogram. The parallelogram is now a rectangle.

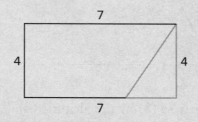

6. What is the area of the rectangle created? Multiply length times width.

7. What is the area of the original parallelogram? Multiply length times altitude.

8. How do the two areas compare?

Something to think about . . .

This rectangle is 7 squares long and 4 squares tall. The total area is 28 square units. The area of the parallelogram is also 28 square units.

BRAIN TICKLERS   Set # 54

Use this diagram to solve problems 1 and 2.

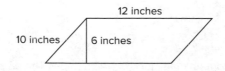

1. What is the perimeter of this parallelogram?
2. What is the area of this parallelogram?
3. What is the area of a parallelogram with base 6 and height 4?

(Answers are on page 231.)

Trapezoids

A *trapezoid* is a quadrilateral with only two parallel sides.

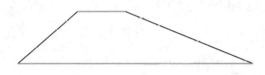

Perimeter

To find the perimeter of a trapezoid, just add all four sides.

Perimeter of a Trapezoid = Base 1 + Base 2 + Side 1 + Side 2

Example:

Find the perimeter of this trapezoid.

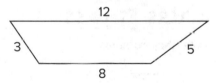

The perimeter of this trapezoid is $3 + 5 + 8 + 12 = 28$ units.

Area

Area of a Trapezoid = $\frac{1}{2}$ (sum of bases) × a

To find the area of a trapezoid, follow these three painless steps:

Step 1: Add base 1 and base 2.

Step 2: Take $\frac{1}{2}$ of the sum of the bases.

Step 3: Multiply the result of Step 2 by the altitude.

$$\text{Area of a trapezoid} = \frac{1}{2}(b_1 + b_2)a$$

Example:

Find the area of this trapezoid.

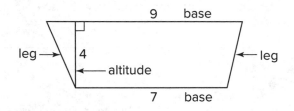

Step 1: First add base 1 to base 2.

$$7 + 9 = 16$$

Step 2: Take $\frac{1}{2}$ of the sum of the bases.

$$\frac{1}{2}(16) = 8$$

Step 3: Multiply the result of Step 2 by the altitude.

$$8(4) = 32$$

The area of the trapezoid is 32 square units.

BRAIN TICKLERS Set # 55

Use the diagrams to solve the problems.

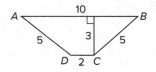

1. What is the perimeter of trapezoid *ABCD*?

2. What is the area of trapezoid *ABCD*?

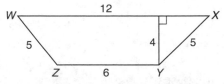

3. What is the perimeter of trapezoid *WXYZ*?

4. What is the area of trapezoid *WXYZ*?

(Answers are on page 231.)

Rhombuses

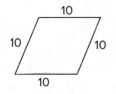

A *rhombus* is a parallelogram with four congruent sides. A square is a special rhombus since a square is a parallelogram with four congruent sides, but a square also has four right angles.

Perimeter

To find the perimeter of a rhombus, add all the sides.

Perimeter of a Rhombus = Side 1 + Side 2 + Side 3 + Side 4

Example:

Find the perimeter of a rhombus if each of the sides of the rhombus is 10 inches.

The perimeter is 10 + 10 + 10 + 10 or 40 inches.

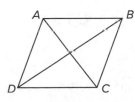

Area

To find the area of a rhombus, follow these two simple steps:

Step 1: Multiply the length of the diagonals together.
$\overline{AC}$ is a diagonal of the rhombus.
$\overline{BD}$ is a diagonal of the rhombus.

Step 2: Take one half of the answer.

Area of a Rhombus = $\frac{1}{2}$ (Diagonal 1 × Diagonal 2)

Example:

If one diagonal of a rhombus is 6 units long and the other diagonal is 5 units long, what is the area of the rhombus?

$$\text{Area} = \frac{1}{2}\,(6 \times 5) = \frac{1}{2}\,(30) = 15 \text{ square units}$$

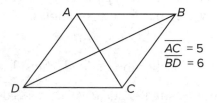

$\overline{AC} = 5$
$\overline{BD} = 6$

EXPERIMENT

Transform a rhombus into another common shape.

Materials
Scissors
Paper
Pencil
Ruler

Procedure
1. Draw a rhombus. Draw the diagonals of the rhombus.

2. Cut the rhombus along its diagonals. Four small triangles are formed.

3. Rearrange the four small triangles into a new shape. What shape can you make?

Something to think about . . .
Is the area of a rhombus related to the area of any other quadrilateral?

BRAIN TICKLERS   Set # 56

1. What is the perimeter of a rhombus if each of the sides is 4 inches long?

2. What is the area of a rhombus with diagonals 6 and 8 inches long?

3. What is the area of a rhombus with diagonals 1 and 2 feet long?

(Answers are on page 232.)

Regular Polygons

A *regular polygon* is a polygon with equal sides and equal angles. An equilateral triangle and a square are both regular polygons. It is possible to construct a regular polygon with any number of sides.

Depending on the number of sides, polygons have different names.

- Three sides = Triangle
- Four sides = Quadrilateral
- Five sides = Pentagon
- Six sides = Hexagon
- Seven sides = Heptagon
- Eight sides = Octagon
- Nine sides = Nonagon

Angles of a polygon

The number of sides of a polygon determines the number of interior angles. A polygon has the same number of sides as interior angles.

The sum of the interior angles of a polygon is $(n - 2)180$.

- The sum of the interior angles of a triangle is $(3 - 2)180 = 180$ degrees.
- The sum of the interior angles of a square is $(4 - 2)180 = 360$ degrees.
- The sum of the interior angles of a pentagon is $(5 - 2)180 = 540$ degrees.
- The sum of the interior angles of a hexagon is $(6 - 2)180 = 720$ degrees.

To illustrate how this equation was determined, divide any polygon into triangles. The sum of the angles in each of the triangles formed is 180 degrees. Multiply the number of triangles by 180 to find the sum of the angles in a polygon.

Divide the square on the following page into two triangles by connecting vertices A and C. The sum of the angles of the two triangles is $180 + 180$ or 360 degrees.

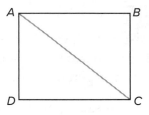

Divide a pentagon into three triangles by connecting point A to points C and D. The sum of the angles of the pentagon will be $180 + 180 + 180$, which is 540 degrees.

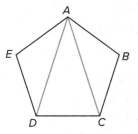

Divide an octagon into six triangles by connecting point A to points $C, D, E, F,$ and G. The sum of the angles of an octagon is $6(180)$, which is 1,080 degrees.

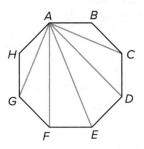

Exterior angles of a polygon

The sum of the exterior angles of a polygon is always 360 degrees.

- The sum of the exterior angles of a triangle is 360 degrees.
- The sum of the exterior angles of a decagon is 360 degrees.
- The sum of the exterior angles of a polygon with 100 sides is still 360 degrees.

BRAIN TICKLERS   Set # 57

1. What is the sum of the interior angles of a pentagon?

2. What is the sum of the exterior angles of a pentagon?

3. What is the sum of the interior angles of a heptagon?

4. What is the sum of the exterior angles of a heptagon?

5. For what shape is the sum of its interior angles equal to its exterior angles?

(Answers are on page 232.)

Perimeter

To find the perimeter of a regular polygon, just add the sides. Or multiply the length of one side by the number of sides. Remember, in a regular polygon, all the sides are the same length.

Perimeter of a Regular Polygon = $n(s)$, where n is the number of sides and s is the length of one side

Example:

Find the perimeter of a hexagon with side 2.

Add the sides of the hexagon together.
$2 + 2 + 2 + 2 + 2 + 2 = 12$
Or multiply the length of one side by the number of sides.
$(6)2 = 12$

Area

Finding the area of a regular polygon is easy.

Multiply $\frac{1}{2}$ the apothem by the perimeter.

The *apothem* is a line segment from the center of the polygon perpendicular to a side.

Apothem

Area of a Regular Polygon $= \frac{1}{2}ap$, where a is the apothem and p is the perimeter

To find the area of a regular polygon, follow these painless steps:

Step 1: Find the length of the apothem.

Step 2: Find the perimeter of the regular polygon.

Step 3: Multiply the apothem by the perimeter.

Step 4: Multiply the answer by $\frac{1}{2}$. The result is the area.

Example:

Find the area of a regular hexagon with side 6 inches and apothem $3\sqrt{3}$ inches.

Step 1:   Find the length of the apothem. The apothem is $3\sqrt{3}$ inches.

Step 2:   Find the perimeter of the regular polygon. The perimeter is 6(6) or 36.

Step 3:   Multiply the apothem by the perimeter.
$$3\sqrt{3}\,(36) = (108)\sqrt{3}$$

Step 4:   Multiply the answer by $\frac{1}{2}$.
$$\frac{1}{2}(108)\sqrt{3} = 54\sqrt{3}$$

The area of the hexagon is $54\sqrt{3}$ square inches.

EXPERIMENT

Compare two methods for finding the area of a square.

Materials
Pencil
Graph paper

Procedure
1. Draw a square with side 6.

2. Draw an apothem. Draw a line segment from the center of the square perpendicular to the opposite side.

3. Measure the length of the apothem.

4. Find the perimeter of the square.

5. Multiply $\frac{1}{2}$ of the apothem of the square by the perimeter to find the area of the square.

6. Now find the area of the same square by multiplying s times s.

7. Compare the areas of the square that you found by the two different methods.

Something to think about . . .
What is the formula to find the perimeter and area of a regular octagon? An octagon is an eight-sided figure.

BRAIN TICKLERS   Set # 58

1. Find the perimeter of an octagon with sides 2 inches long.

2. Find the perimeter of a pentagon with sides 3 feet long.

3. Find the area of an octagon with sides 4 meters and an apothem 3 meters.

4. Find the area of a decagon with sides 3 inches long and apothem 5 inches. A decagon is a 10-sided figure.

(Answers are on page 232.)

SUPER BRAIN TICKLERS #1

Look at each pair of figures. Determine which perimeter is larger or if they are both equal.

1. A rhombus with sides 4 feet long. A pentagon with sides 4 feet long.

2. A parallelogram with length 5 inches and width 10 inches. An equilateral triangle with sides 12 inches long.

3. A trapezoid with bases 4 and 8 centimeters and legs 3 and 9 centimeters. A square with sides 6 centimeters long.

Look at each pair of figures. Determine which area is greater or if they are both the same.

4. A rhombus with diagonals 4 and 8 feet long. A triangle with base 4 feet long and height 8 feet long.

5. A rectangle with sides 3 and 5 miles long. A square with sides 4 miles long.

6. A parallelogram with base 8 kilometers and height 4 kilometers long. A square with sides 6 kilometers long.

(Answers are on page 232.)

PAINLESS TIP

Write the names of each of the following shapes on a separate index card. Write the formula used to compute the area of each shape on the back of each card. Learn each of the following formulas using the flash cards.

Shape	Area	Formula
Triangle	$\frac{1}{2}$ base × height	$\frac{1}{2} bh$
Square	side × side	s^2
Rectangle	length × width	lw
Parallelogram	length × height	lh
Trapezoid	$\frac{1}{2}$ (base$_1$ + base$_2$) × height	$\frac{1}{2} (b_1 + b_2)h$
Rhombus	$\frac{1}{2}$ diagonal$_1$ × diagonal$_2$	$\frac{1}{2} d_1 d_2$
Regular polygon	$\frac{1}{2}$ apothem × perimeter	$\frac{1}{2} ap$

⚡ CAUTION—Major Mistake Territory!

The area of any shape is always in square units. If you write the answer in linear units, it will be incorrect. The area of a square with a side 3 inches is 9 square inches, *not* 9 inches.

Unusual Shapes

To find the area of an unusual shape, add line segments to divide the shape into smaller known shapes. Find the area of each of these smaller shapes and add the result.

Example:

Find the area of the shape on page 215. All the angles are right angles.

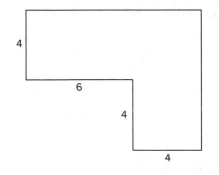

Add a line segment to change the shape into two rectangles.

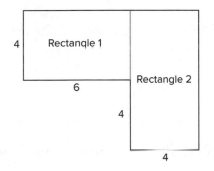

Find the area of rectangle 1 and rectangle 2.

Rectangle 1 has a length of 6 and a width of 4. The area
   of rectangle 1 is 24 square units.
   Rectangle 2 has a length of 4 and a width of 8. The area
      of rectangle 2 is 32 square units.

Add the areas of rectangle 1 and rectangle 2 together to find the
areas of the entire shape. The area of the entire shape is 24 + 32
or 56 square units.

Example:

Find the area of this shape.

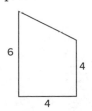

Draw a line segment to divide this shape into a triangle and a square.

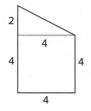

Find the area of the triangle.

$$\text{Area of triangle} = \frac{1}{2} (\text{Base}) \times (\text{Height})$$

The length of the height is 2 units and the length of the base is 4 units. The area of the triangle is $\frac{1}{2}(2)(4) = 4$ square units.

Find the area of the square.

$$\text{Area of square} = \text{Side} \times \text{Side}$$

The length of a side of the square is 4 units. The area is 16 square units.

Add the area of the square and the area of the triangle together to find the area of the shape.

$$4 + 16 = 20 \text{ square units}$$

The area of the shape is 20 square units.

BRAIN TICKLERS   Set # 59

Find the perimeter and area of this shape.

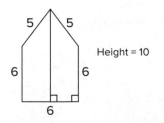

(Answers are on page 232.)

Volume

The volume of a solid figure is the capacity of the figure. The volume is the number of cubic units it contains. Cubes, boxes, cones, balls, and cylinders are all three-dimensional shapes. You can measure the volume of any of these shapes. The volume of three-dimensional shapes is measured in cubic units, such as cubic inches, cubic feet, cubic yards, cubic miles, cubic centimeters, cubic meters, or cubic kilometers. Cubic units are written by placing a small 3 over the units.

EXPERIMENT

Construct a cubic inch.

Materials
Paper
Pencil
Ruler
Scissors
Scotch tape

Procedure
1. Copy the following diagram on a piece of paper. It is made out of six identical squares. Make each square 1 inch by 1 inch.

2. How many right angles are in the diagram?

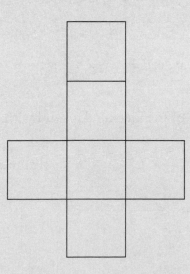

3. Cut out the diagram along the outside border.

4. Fold the diagram along the other edges and make a cube.

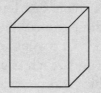

5. Tape the cube into place. This is 1 cubic inch.

6. Count the number of right angles on the cube.

Something to think about . . .
How many inches are in a foot?
How many cubic inches are in a cubic foot?

Rectangular Solids

Rectangular solids are everywhere. A book is a rectangular solid; so is a drawer, a cereal box, a shoebox, a videotape, and a CD case. To find the volume of a rectangular solid, multiply the length times the width times the height of the solid.

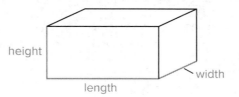

Volume of a Rectangular Solid = $l \times w \times h$ or lwh

Example:

Find the volume of the rectangular solid with length 6, height 4, and width 4.

$$4 \times 6 \times 4 = 96 \text{ cubic units}$$

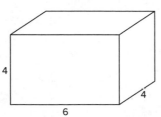

Cubes

A *cube* is a special type of a rectangular solid. The length, width, and height of a cube are exactly the same.

Volume of a Cube $= s \times s \times s = s^3$

Example:

Find the volume of a cube with side 3 inches.

$3 \times 3 \times 3 = 27$ cubic inches
Notice that the answer is in cubic inches.

Surface area

Surface area of a rectangular solid

The surface area is the area on the outside of a three-dimensional shape. Imagine if you had to cover the entire outside of a three-dimensional shape with a piece of paper: how large would the piece of paper be? How could you compute the surface area of a three-dimensional shape?

The surface area of a cube is six times the surface area of one side of the cube. Count the sides of a cube. There are six of them. The surface area of the cube is $6(s)(s)$.

Surface Area of a Cube $= 6s^2$

Example:

What is the surface area of a cube with each side measuring 4 units?

$6(4)(4) = 96$ square units

EXPERIMENT

Find the surface area of a three-dimensional shape.

Materials
Empty cereal box
Ruler
Scissors
Paper
Pencil

Procedure

1. Cut an empty cereal box along the edges of the box. You should cut the cereal box into six pieces. Each side of the box should be a separate piece.

2. Each of the pieces will be a rectangle. Measure the length and width of each rectangle. Enter the results in the chart.

3. Find the area of each of these rectangles by multiplying the length of each rectangle by the width of each rectangle. Enter the results in the chart.

Pieces of box	Length	Width	Area
Front of box			
Back of box			
Top of box			
Bottom of box			
Left side of box			
Right side of box			
Total surface area			

4. Add the areas of all six sides to find the total surface area of a cereal box.

Something to think about . . .

Do any of the sides have the same area?

How would you find the surface area of a pyramid?

To find the surface area of a rectangular solid, just follow these painless steps:

Step 1: Find the length, width, and height of the solid.

Step 2: Multiply 2 × length × width.

Step 3: Multiply 2 × width × height.

Step 4: Multiply 2 × length × height.

Step 5: Add the results of Steps 2, 3, and 4. The answer is the surface area of the rectangular solid.

Example:

Find the surface area of a rectangular solid with sides 5 inches, 6 inches, and 7 inches.

Step 1: Find the length, width, and height of the solid. The length is 5 inches, the width 6 inches, and the height 7 inches.

Step 2: Multiply 2 × length × width.
$2(5)(6) = 60$ square inches

Step 3: Multiply 2 × width × height.
$2(6)(7) = 84$ square inches

Step 4: Multiply 2 × length × height.
$2(5)(7) = 70$ square inches

Step 5: Add the results of Steps 2, 3, and 4. The answer is the surface area of the rectangular solid.
The surface area is 214 square inches.

Surface Area of a Rectangular Solid = $2lw + 2lh + 2hw$

CAUTION—Major Mistake Territory!

Surface area is measured in square units, not units. The surface area of a cube with each side measuring 3 inches is 54 square inches, *not* 54 inches.

BRAIN TICKLERS   Set # 60

1. What is the volume of a cube with side 5 inches?

2. What is the volume of a cube with side 1 inch?

3. What is the surface area of a cube with side 5 inches?

4. What is the surface area of a cube with side 1 inch?

5. What is the volume of a rectangular solid with sides 1, 2, and 3 inches?

6. What is the surface area of a rectangular solid with sides 1, 2, and 3 inches?

(Answers are on page 233.)

Cylinders

A *cylinder* is a common shape. A container of oatmeal is a cylinder, and so is a glass, a can of green beans, or a can of tuna.

To find the surface area of a cylinder, follow these painless steps:

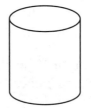

Step 1: Find the height of the cylinder.

Step 2: Find the radius of the cylinder.

Step 3: Find the area of the circle that is the base of the cylinder using the equation $A = \pi r^2$.

Step 4: Multiply the area found in Step 3 by 2 since there is a circle at both the top and the bottom of the cylinder.

Step 5: Find the circumference of the circle that forms the base of the cylinder. Use the formula $2\pi r$.

Step 6: Multiply the circumference of the circle by the height of the cylinder.

Step 7: Add the answers to Steps 4 and 6.

Surface Area of a Cylinder = $2\pi r^2 + 2\pi rh$

Example:

Find the surface area of a cylinder that is 12 inches high and has a diameter of 4 inches.

Step 1: Find the height of the cylinder.

The height is given as 12 inches.

Step 2: Find the radius of the cylinder.
Since the diameter of the cylinder is 4 inches, the radius of the cylinder is 2 inches.

Step 3: Find the area of the circle that is the base of the cylinder using the equation $A = \pi r^2$.
$$A = \pi(2)^2$$
$$A = 4\pi$$

Step 4: Multiply the area found in Step 3 by 2 since there is a circle at both the top and the bottom of the cylinder.
Area of both circles $= 2(4\pi) = 8\pi$

Step 5: Find the circumference of the circle that forms the base of the cylinder. Use the formula $2\pi r$.
Circumference $= 2\pi(2) = 4\pi$

Step 6: Multiply the circumference of the circle (found in Step 5) by the height of the cylinder.
Area of sides of cylinder $= (4\pi)(12) = 48\pi$

Step 7: Add the answer to Step 4 to the answer to Step 6.
$$8\pi + 48\pi = 56\pi$$

The surface area of the cylinder is 56π square inches.

⚡ CAUTION—Major Mistake Territory!

Think of the surface area of a cylinder as the area of the circle on the top of the cylinder plus the area of the circle on the bottom of the cylinder plus the area of the sides of the cylinder. All three areas must be included to get the correct area.

Lateral area of a cylinder

If a cylinder is pictured as a soda can, the lateral area of a cylinder is the curved portion of the can that is printed on. If you could peel the printed section off a soda can, the result would be a rectangle. The length of the rectangle is the circumference of the can. The height of the rectangle is the height of the can.

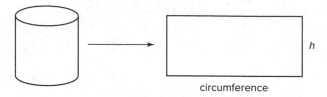

circumference

To find the lateral area of a cylinder, follow these painless steps:

Step 1: Find the height of the cylinder.

Step 2: Find the radius of the cylinder.

Step 3: Use the radius to find the length of the rectangle, which is the same as the circumference of the cylinder. The circumference of the cylinder is $2\pi r$.

Step 4: Multiply the circumference of the cylinder by the height of the cylinder to find the lateral area.

Lateral Area of a Cylinder = $2\pi rh$

Example:

Find the lateral surface of a cylinder with height 7 and diameter 10.

Step 1: Find the height of the cylinder.
The height of the cylinder is 7.

Step 2: Find the radius of the cylinder.
The radius of the cylinder is half the diameter. The radius of the cylinder is $10 \div 2 = 5$.

Step 3: Find the circumference of the cylinder.
The circumference of the cylinder is $2\pi r = 10\pi$.

Step 4: Multiply the circumference of the cylinder by the height of the cylinder to find the lateral area.
The lateral area is $(10\pi)7 = 70\pi$.

To find the volume of a cylinder, follow these three painless steps:

Step 1: Square the radius.

Step 2: Multiply the result by π, which is 3.14.

Step 3: Multiply the answer by the height.

Volume of a Cylinder $= \pi r^2 h$

> **Please add "Volume of a Cylinder = [pi symbol]r2h" between "Step 3:"** and "Example:" (scc ms p. 281) Example:
> Find the volume of a cylinder with radius 5 and height 10.

Step 1: Square the radius.

$$5 \times 5 = 25$$

Step 2: Multiply 25 by π.

$$25\pi$$

Step 3: Multiply 25π by 10.

$$25\pi \times 10 = 250\pi$$

The volume of the cylinder is 250π cubic units.

EXPERIMENT

Compare the volume of three glasses.

Materials
Three glasses of three different sizes
Water
Pencil
Paper
Ruler
Calculator

Procedure

1. Find three glasses of three different sizes.

2. Using the ruler, measure the radius of each glass.

3. Using the ruler, measure the height of each glass.

4. Compute the volume of each glass using the formula $\pi r^2 h$.

Circle	Radius	Height	Volume
Circle 1			
Circle 2			
Circle 3			

5. Based on the volume you computed, rank order the glasses from smallest to largest.

6. Fill what you computed to be the smallest glass with water.

7. Pour the smallest glass of water into the next largest glass. Did all the water fit?

8. Fill this middle-size glass with water and pour it into the largest glass. Did all the water fit? Were all your calculations correct?

Something to think about . . .
How would you compare the volume of a box of cereal to a glass of water?

BRAIN TICKLERS   Set # 61

1. Find the volume of a cylinder with radius 4 and height 2.

2. Find the volume of a cylinder with radius 1 and height 10.

3. Find the volume of a cylinder with radius 10 and height 1.

(Answers are on page 233.)

Cones

The volume of a *cone* is $\frac{1}{3} \pi r^2 h$.

To find the volume of a cone, follow these painless steps:

Step 1: Square the radius.

Step 2: Multiply it by π.

Step 3: Multiply the answer in Step 2 by the height.

Step 4: Multiply the answer in Step 3 by $\frac{1}{3}$.

Notice that the formula for the volume of a cone is exactly one-third the volume of a cylinder of the same height.

Volume of a Cone $= \frac{1}{3}\pi r^2 h$

Example:

Find the volume of a cone with height 5 and diameter 6.

Step 1: Square the radius.

The diameter of the cone is 6.
The radius of the cone is half the diameter, or 3.
3^2 is 9.

Step 2: Multiply 9 by π.

$$9\pi$$

Step 3: Multiply 9π by the height of the cone, which is 5.

$$9\pi \times 5 = 45\pi$$

Step 4: Multiply 45π by $\frac{1}{3}$.

$$45\pi \times \frac{1}{3} = 15\pi$$

The volume of the cone is 15π cubic units.

BRAIN TICKLERS   Set # 62

1. Find the volume of a cone with radius 6 and height 3.

2. Which has a greater volume, a cone with radius 2 and height 6 or a cylinder with radius 2 and height 2?

(Answers are on page 233.)

Spheres

A ball, an orange, and a globe are all spheres. A *sphere* is the set of all points equidistant from a given point.

To find the surface area of a sphere, just follow these painless steps:

Step 1: Find the radius of the sphere.

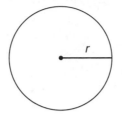

Step 2: Square the radius.

Step 3: Multiply the answer by 4.

Step 4: Multiply the answer by π. The answer is in square units.

Surface Area of a Sphere = $4\pi r^2$

Example:

Find the surface area of a sphere with radius 10 inches.

Step 1: Find the radius of the sphere.
The radius is given as 10.

Step 2: Square the radius.
$$10^2 = 100$$

Step 3: Multiply the answer by 4.
$$4(100) = 400$$

Step 4: Multiply the answer by π. The answer is in square units.
The surface area of the sphere is 400π in^2

All you need to know to find the volume of a sphere is the radius.

To find the volume of a sphere, follow these three painless steps:

Step 1: Cube the radius ($r \times r \times r$).

Step 2: Multiply the answer by $\frac{4}{3}$.

Step 3: Multiply the answer by π.

Volume of a Sphere $= \frac{4}{3}\pi r^3$

Example:

Find the volume of a sphere with radius 6.

Step 1: Cube the radius.

$$6 \times 6 \times 6 = 216$$

Step 2: Multiply the answer by $\frac{4}{3}$.

$$\frac{4}{3} \times 216 = 288$$

Step 3: Multiply the answer by π.

The volume of the sphere is 288π units^3.
If you don't want the answer in terms of π,
multiply 288×3.14.
$288 \times 3.14 = 904.32$ cubic units

BRAIN TICKLERS   Set # 63

1. Compute the surface area of a sphere with radius 3.

2. Compute the volume of a sphere with radius 3.

3. Compute the surface area of a sphere with radius 1.

4. Compute the volume of a sphere with radius 1.

(Answers are on page 233.)

SUPER BRAIN TICKLERS #2

Find the volume of these shapes. Use these formulas.

$$\text{Cube} = s^3$$

$$\text{Rectangular solid} = l \times w \times h$$

$$\text{Cylinder} = \pi r^2 h$$

$$\text{Cone} = \frac{1}{3}\pi r^2 h$$

$$\text{Sphere} = \frac{4}{3}\pi r^3$$

1. A sphere with radius 6.

2. A cone with height 4 and radius 2.

3. A sphere with diameter 10.

4. A cylinder with radius 4 and height 10.

5. A cube with side 8.

6. A rectangular solid with length 1, width 2, and height 3.

(Answers are on page 233.)

BRAIN TICKLERS—THE ANSWERS

Set # 51, page 197

1. 15 inches

2. 9 inches

3. 2 square inches

Set # 52, page 199

1. 32 inches

2. 40 meters

3. 50 square inches

4. 8 square centimeters

Set # 53, page 202

1. 28 units

2. 25 square units

3. 9 square units

4. 40 units

Set # 54, page 204

1. 44 inches

2. 72 square inches

3. 24 square units

Set # 55, page 206

1. 22 units

2. 18 square units

3. 28 units

4. 36 square units

Set # 56, page 208

1. 16 inches

2. 24 square inches

3. 1 square foot

Set # 57, page 211

1. 540

2. 360

3. 900

4. 360

5. A square, rectangle, rhombus, or parallelogram

Set # 58, page 213

1. 16 inches

2. 15 feet

3. 48 square meters

4. 75 square inches

Super Brain Ticklers # 1, page 213

1. A pentagon with sides 4 feet long is larger.

2. An equilateral triangle with sides 12 inches long is larger.

3. They are both the same.

4. They are both the same.

5. A square with sides 4 miles long is larger.

6. A square with sides 6 kilometers long is larger.

Set # 59, page 216

Perimeter = 28 units, Area = 48 square units

Set # 60, page 221

1. 125 cubic inches

2. 1 cubic inch

3. 150 square inches

4. 6 square inches

5. 6 cubic inches

6. 22 square inches

Set # 61, page 226

1. 32π cubic units

2. 10π cubic units

3. 100π cubic units

Set # 62, page 227

1. 36π cubic units

2. They are the same.

Set # 63, page 229

1. 36π square units

2. 36π cubic units

3. 4π square units

4. $\frac{4}{3}\pi$ cubic units

Super Brain Ticklers # 2, page 230

1. 288π cubic units or 904.32 cubic units

2. $5\frac{1}{3}\pi$ cubic units or 16.74 cubic units

3. $1333\frac{1}{3}\pi$ cubic units or 4186.69 cubic units

4. 160π cubic units or 502.4 cubic units

5. 512 cubic units

6. 6 cubic units

Chapter 10

Graphing

Points, lines, polygons, circles, and a variety of other figures can be graphed on a plane. To graph these figures, a grid is drawn. The grid, called a *coordinate plane*, shows where different points are located on the plane. Start by drawing a horizontal number line called the *x*-axis. Draw a vertical number line, called the *y*-axis, perpendicular to the *x*-axis.

Notice:

The *x*-axis and *y*-axis intersect at the origin.
Every point on the *x*-axis to the right of the *y*-axis is positive.
Every point on the *x*-axis to the left of the *y*-axis is negative.
Every point on the *y*-axis above the *x*-axis is positive.
Every point on the *y*-axis below the *x*-axis is negative.

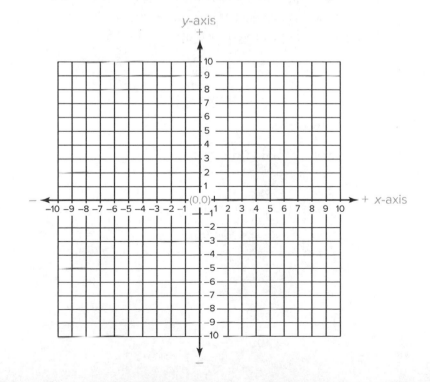

Graphing Points

Coordinate points are used to graph points on the coordinate plane. Coordinate points are written by placing two numbers, x and y, in parentheses, (x, y).

The first number, x, tells where the point is on the x-axis. Positive numbers are located to the right of the y-axis while negative numbers are located to the left of the y-axis. The second number, y, indicates where the point lies on the y-axis. Positive numbers move up above the x-axis, and negative numbers move down below the x-axis.

To graph a coordinate point, follow these four painless steps:

Step 1: Put your pencil at the origin.

Step 2: Look at the x-coordinate. If the number is negative, move the pencil to the left the same number of spaces as the x-coordinate. If the number is positive, move the pencil to the right the same number of spaces as the x-coordinate.

Step 3: Look at the y-coordinate, and move your pencil the same number of spaces as the y-coordinate. If the y-coordinate is negative move down, but if the y-coordinate is positive move up.

Step 4: Put a dot at this point.

Example:
Graph the point $(3, 1)$.

Step 1: Put your pencil at the origin.

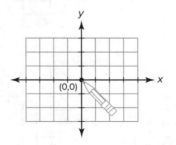

Step 2: Move your pencil right or left along the *x*-axis based on the *x*-coordinate.
Since the *x*-coordinate is 3, move your pencil to the right 3 spaces.

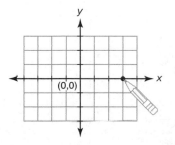

Step 3: Move your pencil up or down based on the *y*-coordinate. Since the *y*-coordinate is 1, move up 1 space.

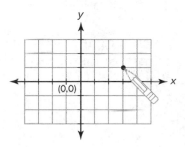

Step 4: Put a dot at this point. Label the point (3, 1).

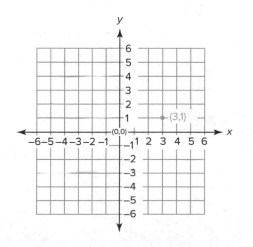

Example:

Graph the point $(0, -3)$.

Step 1: Put your pencil at the origin.

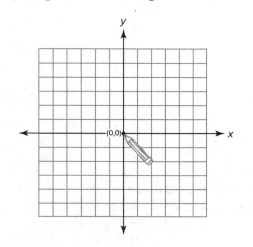

Step 2: Move your pencil right or left along the *x*-axis based on the *x*-coordinate. Since the *x*-coordinate is zero, keep your pencil at the origin.

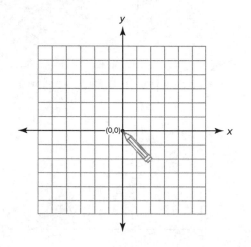

Step 3: Move your pencil up or down based on the *y*-coordinate. Since the *y*-coordinate is −3, move down three spaces.

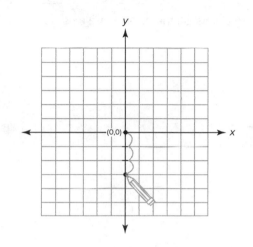

Step 4: Put a dot at this point. Label the point (0, −3).

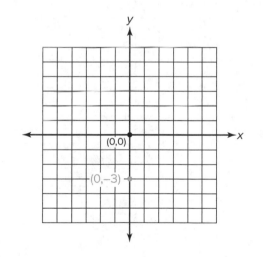

BRAIN TICKLERS   Set # 64

Graph the following points on the coordinate axis below.

1. (1, −3)
2. (−1, 2)
3. (4, 1)
4. (−2, −2)
5. (5, 0)

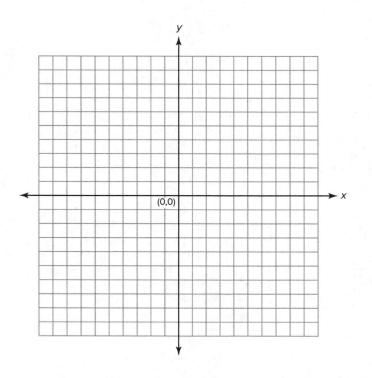

(Answers are on page 272.)

Quadrants

The coordinate plane is divided into four quadrants.

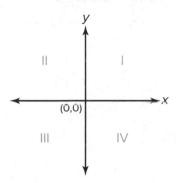

It's possible to determine in which quadrant a point is located just by looking at the coordinates.

Points that contain both a positive x and a positive y are located in quadrant I.

Points that contain a negative x and a positive y are located in quadrant II.

Points that contain both a negative x and a negative y are located in quadrant III.

Points that contain a positive x and a negative y are located in quadrant IV.

⚡ CAUTION—Major Mistake Territory!

The quadrants are counted counterclockwise, not clockwise. Just remember where the positive and negative numbers go and you will have no trouble with quadrants.

Quadrant I points have the form $(+, +)$.
Quadrant II points have the form $(-, +)$.
Quadrant III points have the form $(-, -)$.
Quadrant IV points have the form $(+, -)$.

BRAIN TICKLERS   Set # 65

In which quadrant are the following points located?

1. $(-3, 4)$
2. $(2, 2)$
3. $(100, -4)$
4. $(-2, -2)$
5. $(0, 0)$

(Answers are on page 272.)

The Midpoint Formula

The midpoint of a line segment is the single point equidistant from both endpoints. It's easy to find the midpoint of the segment with the endpoints $(0, 0)$ and $(2, 0)$. Just graph the two points and connect them.

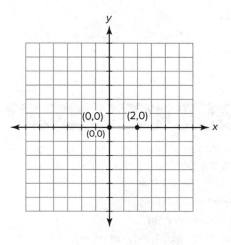

It's obvious that the midpoint is $(1, 0)$ since this segment is just 2 units long. But finding the midpoint of other segments is not so easy. You have to use the midpoint formula.

PAINLESS TIP

Midpoint Formula

The midpoint of the line segment with endpoints (x_1, y_1) and (x_2, y_2) is

$$\left(\frac{x_1 + x_2}{2}, \frac{y_1 + y_2}{2} \right)$$

The midpoint formula uses subscripts, which are the small numbers written after the letters x and y. The small numbers after the x and y are used to tell the difference between the different x's and y's used in the equation.

x_1 is read x sub one.

x_2 is read x sub two.

y_1 is read y sub one.

y_2 is read y sub two.

To find the midpoint of a line segment, just follow these painless steps:

Step 1: Find the endpoints of the line segment.

Step 2: Add the two x-coordinates and divide by 2.

Step 3: Add the two y-coordinates and divide by 2.

Step 4: The answers to Step 2 and Step 3 are the coordinates of the midpoint.

Example:

Find the midpoint of the segment that connects the points $(3, 2)$ and $(1, 6)$.

Step 1: Find the coordinates of the endpoints of the segment. The endpoints are given as $(3, 2)$ and $(1, 6)$.

Step 2: Add the x-coordinates and divide by 2.

$$\frac{(3+1)}{2} = \frac{4}{2} = 2$$

Step 3: Add the y-coordinates and divide by 2.

$$\frac{(2+6)}{2} = \frac{8}{2} = 4$$

Step 4: The coordinates of the midpoint are $(2, 4)$.

Example:

Find the midpoint of the segment that connects the points $(0, -3)$ and $(4, -1)$.

Step 1: Find the coordinates of the endpoints of the segment. The endpoints are given as $(0, -3)$ and $(4, -1)$.

Step 2: Add the x-coordinates and divide by 2.

$$\frac{(0+4)}{2} = \frac{4}{2} = 2$$

Step 3: Add the y-coordinates and divide by 2.

$$\frac{(-3+-1)}{2} = \frac{-4}{2} = -2$$

Step 4: The coordinates of the midpoint are $(2, -2)$.

EXPERIMENT

Find the midpoint of the line segment.

Materials
Graph paper
Pencil
Ruler

Procedure
1. Graph each of the line segments.

2. Estimate the midpoint of each segment by examining the line segment. Enter your estimate in the chart.

3. Determine the midpoint of each segment using the midpoint formula:

$$\left(\frac{x_1 + x_2}{2}, \frac{y_1 + y_2}{2} \right)$$

Enter the results in the chart.

Line segment	Endpoint 1	Endpoint 2	Estimate midpoint	Midpoint formula
Line segment 1	(3, 1)	(−1, 1)		
Line segment 2	(2, 8)	(2, 3)		
Line segment 3	(3, 5)	(1, 3)		
Line segment 4	(3, 3)	(−3, −3)		
Line segment 5	(−1, −4)	(2, 6)		

Something to think about . . .
What is the difference between your estimate and using the midpoint formula?

BRAIN TICKLERS   Set # 66

Find the midpoint of each of the segments formed by connecting the following pairs of points.

1. (6, 4) and (−2, 6)

2. (2, 2) and (8, 8)

3. (−3, −3) and (5, 5)

4. (0, −2) and (4, 0)

(Answers are on page 273.)

The Distance Formula

The length of a line segment is the distance from one end of the segment to the other. Distance is measured in linear units (e.g., inches, feet, meters).

To find the length of a line segment, just follow these painless steps:

Step 1: Find the endpoints of the line segment.

Step 2: Subtract the smaller x-coordinate from the larger x-coordinate.

Step 3: Square the difference.

Step 4: Subtract the smaller y-coordinate from the larger y-coordinate.

Step 5: Square the difference.

Step 6: Add Step 3 and Step 5.

Step 7: Take the square root of Step 6. That's the distance from one end of the segment to the other.

Distance Formula

The distance between any two points (x_1, y_1) and (x_2, y_2) is

$$\sqrt{\left(x_2 - x_1\right)^2 + \left(y_2 - y_1\right)^2}$$

Example:

What is the distance between the points $(5, 5)$ and $(1, 2)$?

Step 1: Find the endpoints of the line segment. The endpoints of the segment are given as $(5, 5)$ and $(1, 2)$.

Step 2: Subtract the smaller x-coordinate from the larger x-coordinate.
$$5 - 1 = 4$$

Step 3: Square the difference.
$$4 \times 4 = 16$$

Step 4: Subtract the smaller y-coordinate from the larger y-coordinate.
$$5 - 2 = 3$$

Step 5: Square the difference.
$$3 \times 3 = 9$$

Step 6: Add Step 3 and Step 5.
$$16 + 9 = 25$$

Step 7: Take the square root of Step 6.

$$\sqrt{25} = 5$$

The distance between the points $(1, 2)$ and $(5, 5)$ is 5.

EXPERIMENT

Find the length of a line segment using the Pythagorean theorem.

Materials
Graph paper
Pencil
Ruler

Procedure

1. Draw a set of coordinate axes on a piece of graph paper.

2. Pick any two points and connect them.

3. Make a right triangle using the two points as two of the vertices of the triangle.

   Draw a line parallel to the x-axis through the point with the lower y value.

   Draw a line parallel to the y-axis through the point with the higher y value. Where the two lines intersect is the right angle of the right triangle.

4. Count the squares to determine the length of each side of the triangle. Enter the results in the chart.

5. Compute the length of the hypotenuse. Use the Pythagorean theorem. Enter the results in the chart.

6. Compute the distance between the two points using the distance formula. Enter the results in the chart.

7. Repeat Steps 1 to 6, picking another set of 2 points.

8. Repeat Steps 1 to 6 until the chart is complete.

Set	Point 1	Point 2	Length of horizontal side	Length of vertical side	Length of hypotenuse	Distance formula
Set 1						
Set 2						
Set 3						
Set 4						

Something to think about . . .
Compare the answers you found using the distance formula to the length of the hypotenuse. Are they different?

⚡ CAUTION—Major Mistake Territory!

The distance between two points is always a positive number.

🧩 BRAIN TICKLERS   Set # 67

Find the distance between these two points.

1. (4, 2) and (1, 6)

2. (0, 6) and (1, 4)

3. (−3, 2) and (4, −1)

4. (−2, 0) and (0, 1)

(Answers are on page 273.)

Graphing a Line by Plotting Points

An equation of a line has x- and/or y-variables. The following equations are lines.

$$x + y = 9$$

$$3x + 2y = 1$$

$$y = -3x + 2$$

$$y = 4x$$

$$y = -5$$

$$x = 7$$

These are *not* equations of a line, since these equations have an x^2 term and/or a y^2 term.

$$3x^2 + 5y = 7$$

$$2x + 9y^2 = 1$$

$$3x^2 - 2y^2 = 4$$

There are several ways to graph a line. The easiest way is plotting points. You can graph a line with only two points, since two points determine a line. But it is best to plot three points to make sure you didn't make a mistake. You can pick any numbers you want. Why not pick numbers that make the calculations easy?

To plot points, just follow these five painless steps:

Step 1: Substitute 0 for x and solve for y.

Step 2: Substitute 1 for x and solve for y.

Step 3: Substitute -1 for x and solve for y.

Step 4: Graph the three points.

Step 5: Connect the three points to form a line. Label the line.

If the three points do not lie in a straight line, you have solved one of the equations incorrectly or graphed one of the three points incorrectly.

Example:
Graph the equation $y = 3x - 2$.

Step 1: Substitute 0 for x and solve for y.
$$y = 3(0) - 2$$
$$y = -2$$
The first point is $(0, -2)$.

Step 2: Substitute 1 for x in the equation and solve for y.
$$y = 3(1) - 2$$
$$y = 1$$
The second point is $(1, 1)$.

Step 3: Substitute -1 for x in the equation and solve for y.
$$y = 3(-1) - 2$$
$$y = -5$$
The third point is $(-1, -5)$.

Step 4: Graph the points.

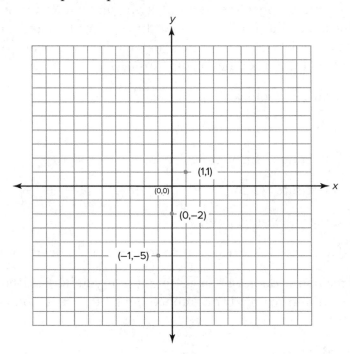

Step 5: Connect the points to form the line $y = 3x - 2$. Label the line.

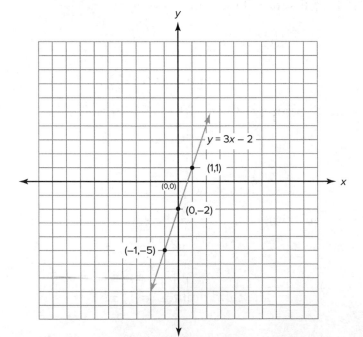

BRAIN TICKLERS   Set # 68

Graph the following lines.

1. $y = 4x - 6$
2. $2x + 2y = 10$
3. $x = y$

(Answers are on page 273.)

Graphing Horizontal and Vertical Lines

Horizontal and vertical lines are exceptions to the graphing rule. Horizontal lines do not have an x term. They are written in the form $y =$ some number. $y = 3$, $y = 0$, and $y = -7$ are all horizontal lines.

Example:

Graph the equation $y = 4$.

Just plot points.
If x is 0, y is 4.
If x is 1, y is 4.
If x is -1, y is 4.
No matter what number x is, y is always 4.
The graph of $y = 4$ is a horizontal line that intersects the y-axis at 4.

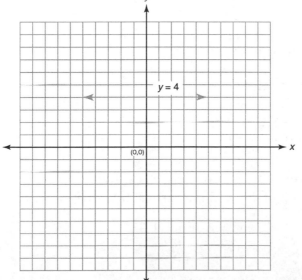

Example:

Graph the line $y = -2$. Just draw a horizontal line at $y = -2$ parallel to the x-axis, since no matter what x is, $y = -2$.

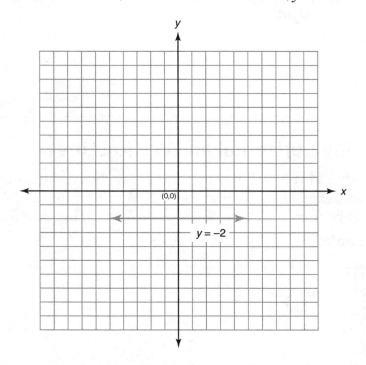

Vertical lines do not have a y term. They are parallel to the y-axis. These lines are parallel to the y-axis: $x = 1$, $x = -5$, and $x = 0$.

Example:

Graph the equation $x = -2$.

Just plot points.
If $y = 1$, $x = -2$.
If $y = -1$, $x = -2$.
If $y = 0$, $x = -2$.

No matter what number y is, x is always -2.
The graph of $x = -2$ is a vertical line that intersects the x-axis at -2.

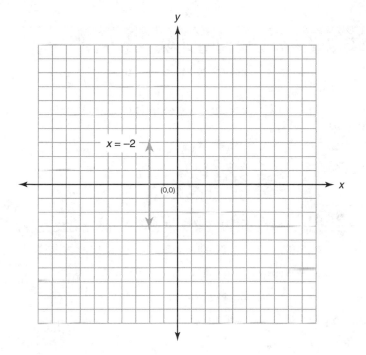

Example:

Graph the equation $x = 5$. Just graph a line parallel to the y-axis at $x = 5$, since no matter what y is, x is always 5.

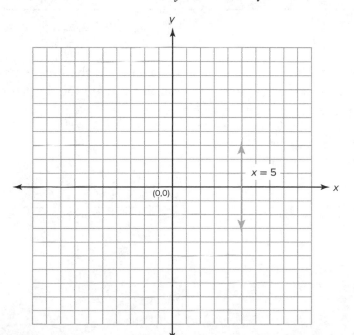

CAUTION—Major Mistake Territory!

Any line of the form y = a number intersects the y-axis. It is not parallel to the y-axis.

Any line of the form x = a number intersects the x-axis. It is not parallel to the x-axis.

BRAIN TICKLERS   Set # 69

Decide whether each of the following statements is true or false.

1. The line $y = 4$ is parallel to the x-axis.

2. The line $x = -1$ intersects the x-axis.

3. The line $y = 3$ intersects the y-axis.

4. The line $y = 0$ is the x-axis.

5. The line $x = 0$ is the x-axis.

6. The line $x = 3$ is parallel to the y-axis.

(Answers are on page 275.)

The Slope

The *slope* of a line is a measure of the incline of the line. A positive slope indicates that the line goes uphill if you are moving from left to right. These lines have positive slopes.

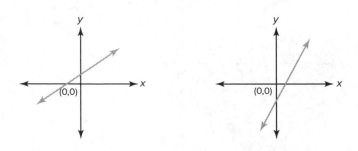

A negative slope indicates that a line goes downhill if you are moving from left to right. These lines have negative slopes.

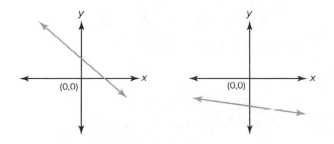

There are two common ways to find the slope of a line.

1. Putting an equation in slope-intercept form.

2. Using the point-point method.

Method 1: Slope-intercept method

To change an equation to slope-intercept form, just follow these two
painless steps:

Step 1: Solve the equation for y.

Step 2: The variable in front of the x is the slope.

Example:

Find the slope of the line $2x + y - 2 = 0$.

Step 1: Solve the equation $2x + y - 2 = 0$ for y.

Subtract $2x$ from both sides of the equation.

$$y - 2 = -2x$$

Add 2 to both sides of the equation.

$$y = -2x + 2$$

Step 2: The number in front of x is the slope.

2 is in front of x, so -2 is the slope.

Example:

Find the slope of the equation $3x + 6y = 18$.

Step 1: Solve the equation $3x + 6y = 18$ for y.

Subtract $3x$ from both sides of the equation.

$$6y = -3x + 18$$

Divide both sides of the equation by 6.

$$y = -\frac{1}{2}x + 3$$

Step 2: The number in front of the x is the slope.

The slope is $-\frac{1}{2}$.

BRAIN TICKLERS   Set # 70

Change each of the following equations to slope-intercept form and find the slope.

1. $5x - y + 2 = 0$

2. $4x + 2y = -8$

3. $7y = -3x$

4. $x + 2y = 1$

(Answers are on page 275.)

Method 2: Point-point method

To find the slope of a line using the point-point method, just find two points on the line.

Using two points to find the slope of a line is painless:

Step 1: Find two points on the line.

Step 2: Subtract the first y-coordinate (y_1) from the second y-coordinate (y_2) to find the change in y.

Step 3: Subtract the first x-coordinate (x_1) from the second x-coordinate (x_2) to find the change in x.

Step 4: Divide the change in y (Step 2) by the change in x (Step 3). The answer is the slope of the line.

$$\text{Slope} = \frac{y_2 - y_1}{x_2 - x_1}$$

Example:

Find the slope of a line through the points $(3, 2)$ and $(7, 1)$.

Step 1: Find two points on the line. Two points on the line are given: $(3, 2)$ and $(7, 1)$.

Step 2: Subtracting the second y-coordinate from the first y-coordinate, find the change in y.

$$2 - 1 = 1$$

Step 3: Subtracting the second x-coordinate from the first x-coordinate, find the change in x.

$$3 - 7 = -4$$

Step 4: Find the slope of the line by dividing the change in y (Step 2) by the change in x (Step 3).

$$\frac{1}{(-4)} = -\frac{1}{4}$$

The slope is $-\frac{1}{4}$.

EXPERIMENT

Estimate the slope of a line.

Materials
 Pencil
 Graph paper
 Ruler

Procedure
 1. Graph the pairs of points shown in the chart.

 2. Connect each pair of points to form a line.

 3. Look at each line and try to estimate the slope. Enter your guess in the chart.

Use these tips for help.

- If the slope goes uphill, it is positive.

- If the slope goes downhill, it is negative.

- The steeper the slope, the higher the number.

- Flatter lines have slopes less than 1.

4. Compute the slope by finding the change in *y* divided by the change in *x*. Enter your results in the chart in the column labeled "Actual slope."

5. Compare your estimates to the actual slopes.

Line	Point 1	Point 2	Estimated slope	Actual slope
Line 1	(1, 1)	(5, 5)		
Line 2	(0, 5)	(5, 0)		
Line 3	(0, 0)	(1, 5)		
Line 4	(−5, 1)	(5, −1)		

Something to think about . . .

What would a line with slope $\frac{1}{10}$ look like compared to a line with slope 10?

BRAIN TICKLERS   Set # 71

Find the slope of the line through the following points.

1. (3, 0) and (0, −2)

2. (7, 1) and (2, 6)

3. (0, 0) and (5, 3)

4. (−3, −2) and (1, 1)

(Answers are on page 275.)

Graphing Using Slope-Intercept

The easiest way to graph a line is to use the slope-intercept method. The words *slope-intercept* refer to the form of the equation. When an equation is in slope-intercept form it is written in terms of a single *y*.

Equations in slope-intercept form have the form $y = mx + b$.

> *y* is a variable.
> *m* is the slope of the line.
> *b* tells you where the line intercepts the *y*-axis.

Follow these six painless steps to graph an equation using the slope-intercept form.

Step 1: Put the equation in slope-intercept form. Express the equation of the line in terms of a single *y*. The equation should have the form $y = mx + b$.

Step 2: The number without any variable after it (the *b* term) is the *y*-intercept. The *y*-intercept is where the line crosses the *y*-axis. Make a mark on the *y*-axis at the *y*-intercept. If the equation has no *b* term, the *y*-intercept is 0.

Step 3: The number in front of the *x* is the slope. In order to graph the line, the slope must be written as a fraction. If the slope is a fraction, leave it as it is. If the slope is a whole number, place it over 1. If the slope is a decimal, convert it to a fraction.

Step 4: Start at the *y*-intercept and move your pencil up the *y*-axis the number of spaces in the numerator of the slope.

Step 5: If the fraction is negative, move your pencil to the left the number of spaces in the denominator of the slope. If the fraction is positive, move your pencil to the right the number of spaces in the denominator of the slope. Mark this point.

Step 6: Construct a line that connects the point on the *y*-axis with the point where the pencil ended up. An equation in slope-intercept form has the form $y = mx + b$, where m is the slope and b is where the line intercepts the *y*-axis.

Example:

Graph the line $2y - 4x = 8$.

Step 1: Put equation $2y - 4x = 8$ in slope-intercept form.

Express the equation in terms of a single y.
$2y - 4x = 8$

Add $4x$ to both sides.
$2y = 4x + 8$

Divide both sides of the equation by 2.
$y = 2x + 4$

This equation is now in slope-intercept form.

Step 2: The number at the end of the equation without any variable after it is the y-intercept. The y-intercept is where the line crosses the y-axis. Make a mark at the y-intercept. The y-intercept is 4.

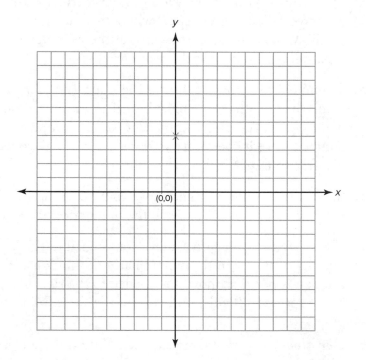

Step 3: The number in front of the x is the slope. If the slope is a whole number, place it over 1. If it is a fraction, leave it as it is. The slope is 2, so change it to $\frac{2}{1}$.

Step 4: Look at the numerator. Start at the y-intercept and move your pencil up the y-axis the number of spaces in the numerator. The numerator is 2, so move up two spaces to the point $y = 6$.

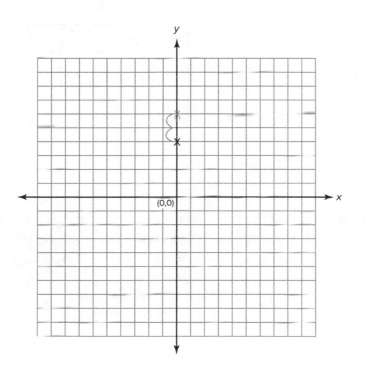

Step 5: Look at the denominator. If the fraction is negative, move your pencil to the left the number of spaces in the denominator. If the fraction is positive, move your pencil to the right the number of spaces in the denominator. The fraction is positive and the denominator is 1, so move your pencil one space to the right.

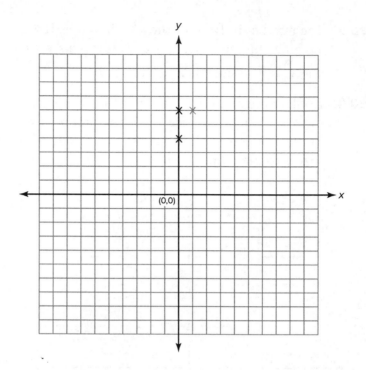

Step 6: Connect the point on the *y*-axis with the point where the pencil ended up and extend both ends to form a line.

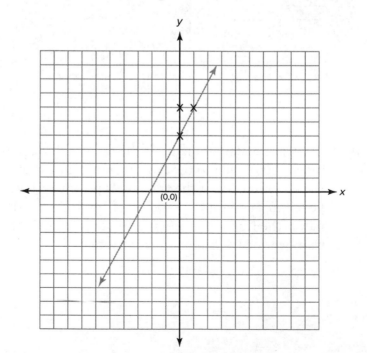

BRAIN TICKLERS   Set # 72

What is the slope and *y*-intercept of each of these lines?

1. $y = 3x - 2$
2. $2y = 2x$
3. $4x - 4y = 8$
4. $2y - x = -10$
5. $y = 3$

(Answers are on page 275.)

Finding the Equation of a Line

If you know the slope of a line and a point on the line, you can find the equation of the line.

Just follow these four painless steps to find the equation of a line.

Step 1: Substitute the slope of the line for m in the equation $y = mx + b$.

Step 2: Substitute the coordinates of a point on the line for the variables x and y in the equation $y = mx + b$.

Step 3: Solve for b.

Step 4: Substitute m and b into the equation $y = mx + b$ to find the equation of a line.

Example:

Find the equation of a line with slope -5 and through the point $(-2, -1)$.

Step 1: Substitute the slope, which is -5, for m in the equation $y = mx + b$.

$$y = -5x + b$$

Step 2: Substitute the coordinates of the point on the line for the variables x and y in the equation $y = mx + b$.

$$(-1) = -5(-2) + b$$

Step 3: Solve for b.

$$(-1) = (-5)(-2) + b$$
$$(-1) = 10 + b$$
$$b = -11$$

Step 4: Substitute m and b into the equation $y = mx + b$ to find the equation of the line. The equation of the line is

$$y = -5x - 11$$

BRAIN TICKLERS   Set # 73

Find the equation of each of these lines.

1. Slope = 2 and point (4, −4)

2. Slope = 1 and point (1, 1)

3. Slope = $\frac{1}{2}$ and point (0, −1)

(Answers are on page 276.)

Two-Point Method

If you know any two points, you can also find the equation of a line.

Just follow these four painless steps to find the equation of a line.

Step 1: Find the slope of the line. Divide the change in y by the change in x.

Step 2: Substitute the slope (m) into the equation $y = mx + b$.

Step 3: Substitute one pair of coordinates into the equation for x and y and solve for b.

Step 4: Substitute m and b into the equation $y = mx + b$ to find the equation of a line.

Example:

Find the equation of the line through the points $(3, 4)$ and $(6, 2)$.

Step 1: Find the slope of the line.

Subtract one set of coordinates from the other set of coordinates.
  The change in y is $4 - 2$ or 2.
  The change in x is $3 - 6$ or -3.
Divide the change in y by the change in x to find the slope.
The slope is $-\dfrac{2}{3}$.

Step 2: Substitute the slope (m) into the equation $y = mx + b$.

$$y = -\frac{2}{3}x + b$$

Step 3: Substitute one pair of coordinates into the equation for x and y and solve for b.

$$4 = -\frac{2}{3}(3) + b$$
$$4 = -2 + b$$
$$b = 6$$

Step 4: Substitute m and b into the equation $y = mx + b$. As a result, the equation of a line through the points $(3, 4)$ and $(6, 2)$ is

$$y = -\frac{2}{3}x + 6$$

BRAIN TICKLERS   Set # 74

Find the equation of the line through each pair of points.

1. $(1, 1)$ and $(6, 2)$

2. $(0, 5)$ and $(2, 0)$

3. $(3, -1)$ and $(4, -2)$

(Answers are on page 276.)

Point-Slope Form

Another way to find the equation of a line is by using the point-slope formula.

Point-slope formula: $y - y_1 = m(x - x_1)$

where x_1 and y_1 are the coordinates of a single point on the line, and m is the slope of the line.

Example:

Find the equation of a line with slope 3 and point $(2, 5)$.
Substitute $x = 2, y = 5$, and $m = 3$ into the point-slope formula.

$$y - 5 = 3(x - 5)$$

Solve.

$$y - 5 = 3x - 15$$
$$y = 3x - 10$$

Example:

Find the equation of a line with slope $\frac{1}{2}$ and passing through the point $(-4, 6)$.

Substitute $x = -4, y = 6$, and $m = \frac{1}{2}$ into the point-slope formula.

$$y - 6 = \frac{1}{2}(x - (-4))$$

Solve.

$$y - 6 = \frac{1}{2}(x + 4)$$

$$y - 6 = \frac{1}{2}x + 2$$

$$y = \frac{1}{2}x + 8$$

BRAIN TICKLERS   Set # 75

1. What is the equation of a line with point (3, 5) and slope −2?

2. What is the equation of a line with point (0, 0) and slope 1?

3. What is the equation of a line with point (−2, −6) and slope −1?

4. What is the equation of a line with point (1, 1) and slope 0?

(Answers are on page 276.)

SUPER BRAIN TICKLERS   # 1

1. What is the equation of a line that goes through the point (5, 5) and has slope −2?

2. What is the equation of a line that goes through the point (−3, 3) and has slope 0?

(Answers are on page 276.)

Parallel and Perpendicular Lines

EXPERIMENT

Learn the relationship between lines with the same slope.

Materials
Graph paper
Pencil
Ruler

Procedure
1. Graph all three of the following equations on the same pair of coordinate axes.

$$y = 2x + 3$$

$$y = 2x - 1$$

$$2y = 4x + 2$$

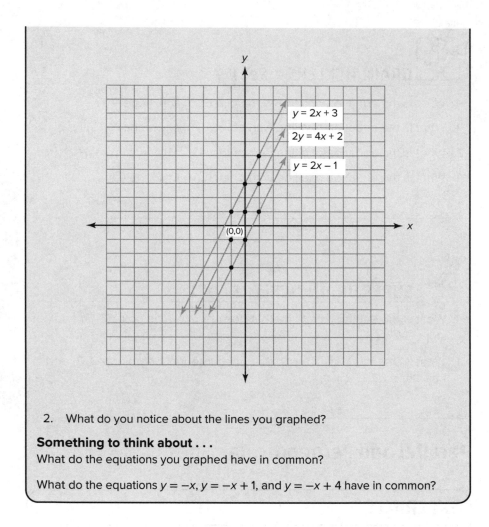

2.   What do you notice about the lines you graphed?

Something to think about . . .

What do the equations you graphed have in common?

What do the equations $y = -x$, $y = -x + 1$, and $y = -x + 4$ have in common?

THEOREM   If two lines are parallel, then they have the same slope.

THEOREM   If two lines have the same slope, then they are parallel.

Example:

$y = 2x + 1$ and $y = 2x - 6$ are parallel since they both have a slope of 2.

EXPERIMENT

Discover the relationship between perpendicular lines.

Materials
Graph paper
Pencil
Ruler

Procedure

1. Graph the following pair of points (0, 0) and (3, 3).

2. Connect the points to form line 1.

3. Graph the following points (3, 0) and (0, 3).

4. Connect the points to form line 2.

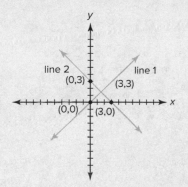

5. Line 1 and line 2 should be perpendicular to each other.

6. Determine the slope of line 1.

7. Determine the slope of line 2.

8. Multiply the two slopes together. What's your answer?

9. On a second set of coordinate axes draw line 1 through the points (0, 1) and (1, 3).

10. Draw line 2 through the points (0, 1) and (2, 0).

11. Line 1 should be perpendicular to line 2.

12. Find the slope of both these lines and multiply them together. What's your answer?

13. Do you notice a pattern?

Something to think about . . .

How can you tell if two lines are perpendicular to each other?

THEOREM   If two lines are perpendicular, then the slope of one line is the negative reciprocal of the slope of the other line.

THEOREM   If the slope of one line is the negative reciprocal of the slope of another line, then the lines are perpendicular.

Example:

$y = -3x + 2$ and $y = \frac{1}{3}x - 1$ are perpendicular lines.

Multiply the slope of the first line (-3) times the slope of the second line $\left(\frac{1}{3}\right)$. The answer is (-1).

BRAIN TICKLERS   Set # 76

Look at the following pairs of equations. Determine whether each pair of lines is perpendicular, parallel, or neither.

1. $y = 3x - 3$ and $y = -\frac{1}{3}x + 3$

2. $y = 2x + 5$ and $y = 2x + 7$

3. $y = -4x + 1$ and $y = -6x - 2$

4. $2y = x + 1$ and $y = \frac{1}{2}x + 3$

(Answers are on page 276.)

PAINLESS TIP

Put the name of each of the following equations on one side of an index card and the formula on the other side. Memorize these formulas!

Midpoint formula

Distance formula

Horizontal line

Vertical line

Slope

Slope-intercept form

Parallel lines

Perpendicular lines

SUPER BRAIN TICKLERS   # 2

1. What quadrant is the point (3, −2) located in?

2. What is the midpoint of the segment with endpoints (3, 2) and (−7, 0)?

3. What is the distance between the points (−1, −1) and (4, 5)?

4. Put the line $2x + 10y = 40$ in slope-intercept form.

5. Find the slope of the line through the points (5, 1) and (−1, −2).

6. Find the equation of the line with slope 4 and intercept (0, 0).

7. At what point does the line $y = 4$ cross the y-axis?

8. At what point does the line $x = 4$ cross the y-axis?

9. What is the equation of the y-axis?

10. What is the equation of the line through the points (3, 5) and (1, 8)?

11. What is the relationship between the lines $y = -\frac{1}{2}x + 1$ and $y = 2x$?

12. What is the relationship between the lines $10y = -5x + 4$ and $y = -\frac{1}{2}x - 1$?

(Answers are on page 277.)

BRAIN TICKLERS—THE ANSWERS

Set # 64, page 240

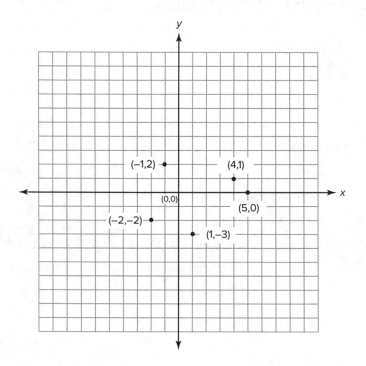

Set # 65, page 242

1. II
2. I
3. IV
4. III
5. The origin

Set # 66, page 245

1. $(2, 5)$

2. $(5, 5)$

3. $(1, 1)$

4. $(2, -1)$

Set # 67, page 248

1. 5 units

2. $\sqrt{5}$ units

3. $\sqrt{58}$ units

4. $\sqrt{5}$ units

Set # 68, page 251

1.

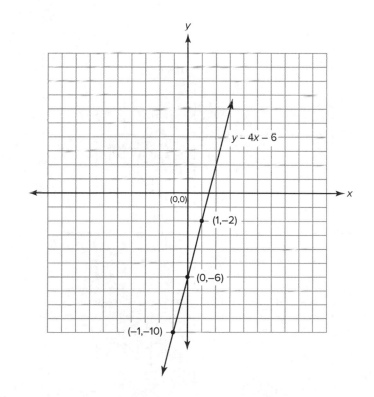

2.

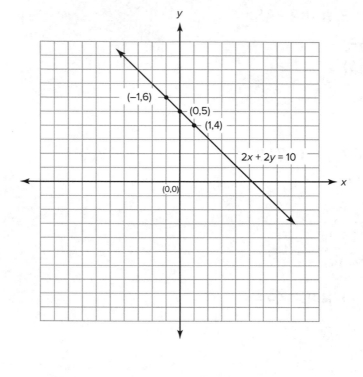

3.

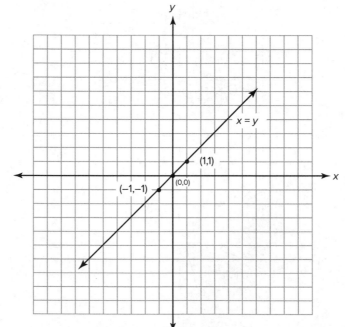

Set # 69, page 254

  1. True

  2. True

  3. True

  4. True

  5. False

  6. True

Set # 70, page 256

  1. $y = 5x + 2$, slope $= 5$

  2. $y = -2x - 4$, slope $= -2$

  3. $y = -\dfrac{3}{7} x$, slope $= \dfrac{3}{7}$

  4. $y = -\dfrac{1}{2} x + \dfrac{1}{2}$, slope $= -\dfrac{1}{2}$

Set # 71, page 258

  1. $\dfrac{2}{3}$

  2. -1

  3. $\dfrac{3}{5}$

  4. $\dfrac{3}{4}$

Set # 72, page 263

  1. Slope $= 3$, y-intercept $= -2$

  2. Slope $= 1$, y-intercept $= 0$

  3. Slope $= 1$, y-intercept $= -2$

  4. Slope $= \dfrac{1}{2}$, y-intercept $= -5$

  5. Slope $= 0$, y-intercept $= 3$

Set # 73, page 264

1. $y = 2x - 12$
2. $y = x$
3. $y = \dfrac{1}{2}x - 1$

Set # 74, page 265

1. $y = \dfrac{1}{5}x + \dfrac{4}{5}$
2. $y = -\dfrac{5}{2}x + 5$
3. $y = -x + 2$

Set # 75, page 267

1. $y = -2x + 11$
2. $y = x$
3. $y = -x - 8$
4. $y = 1$

Super Brain Ticklers # 1, page 267

1. $y = -2x + 15$
2. $y = 3$

Set # 76, page 270

1. Perpendicular
2. Parallel
3. Neither
4. Parallel

Super Brain Ticklers # 2, page 271

1. Quadrant IV

2. $(-2, 1)$

3. $\sqrt{61}$

4. $y = -\dfrac{1}{5}x + 4$

5. $\dfrac{1}{2}$

6. $y = 4x$

7. $(0, 4)$

8. Never

9. $x = 0$

10. $y = -\dfrac{3}{2}x + 9\dfrac{1}{2}$

11. Perpendicular

12. Parallel

Chapter 11

Constructions

The word *constructions* in geometry refers to creating or replicating segments, lines, angles, or shapes accurately with only three pieces of equipment: a compass, a straight edge, and a pencil

A compass is a tool used for constructing arcs and circles. A compass has two arms. One arm has a point at the end, and the other arm has a pencil at the end. The distance between the two arms can be changed to create different-size arcs and circles. A straight edge is used to guide a pencil to make straight lines. Typically, a ruler is used as a straight edge, but a straight edge could be any object with a straight side. The straight edge is not used to measure length or width. It is only used to draw straight lines.

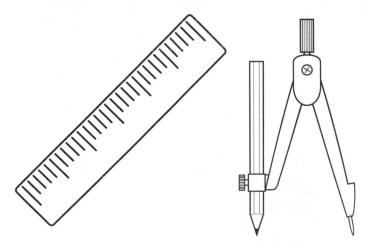

Constructing a Congruent Line Segment

Construct a line segment exactly the same length as an existing line segment with just a straight edge, a compass, and a pencil. It's easy. Just follow these three painless steps.

Copy the line segment *AB*.

Step 1: Use the straight edge and pencil to draw a line segment that is longer than segment *AB*.

Label one end of the segment *C*.

Step 2: Use the compass to measure the length of segment *AB*. Put the point of the compass on point *A*. Put the pencil end of the compass on point *B*.

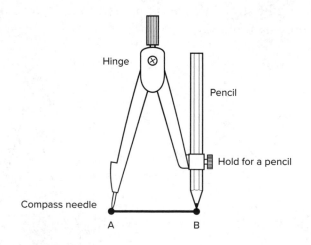

Step 3: Keeping the opening of the compass the same, place the point of the compass on point *C*. Move the pencil end of the compass to create an arc on the line segment. Label the point *D*, where the line segment and arc intersect.

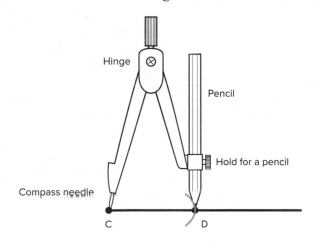

Segment *CD* and segment *AB* are both the same length. They are congruent.

Constructing Congruent Angles

Use a compass, a straight edge, and a pencil to construct an angle congruent to an existing angle. Both angles will have the same measure. It's easy! Just follow these six painless steps.

Construct an angle *FGH* congruent to angle *ABC*.

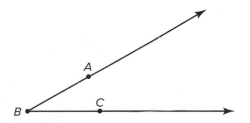

Step 1: Draw a ray. Label the point at the end of the ray *G*.

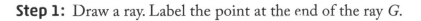

Step 2: Put the point of the compass at point B of the original angle. Construct an arc that intersects both sides of the angle. Label the points of intersection D and E.

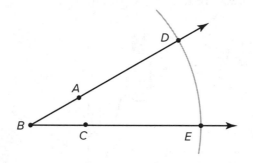

Step 3: Put the point of the compass at point G on the ray. Draw the same sized arc on the ray. Mark the point of intersection H.

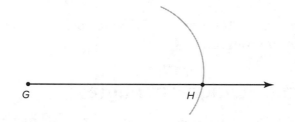

Step 4: On the original angle ABC, use the compass to measure the distance between the two points where the arc intersects the sides of the angle.

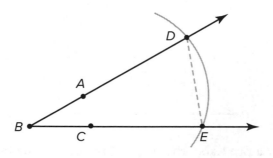

Step 5: Keeping the arms of the compass in the same place, put the point of the compass at point H, and make an arc that intersects the original arc. Mark the point of intersection F.

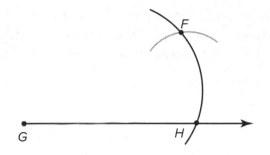

Step 6: Connect point G to the point where the two arcs intersect at point F.

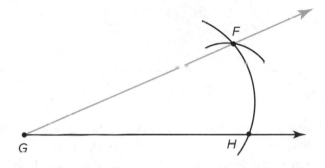

Angle FGH is congruent to angle ABC.

Bisecting an Angle

It is possible to bisect any angle using a compass, a straight edge, and a pencil. Just follow these three painless steps.

Bisect angle ABC.

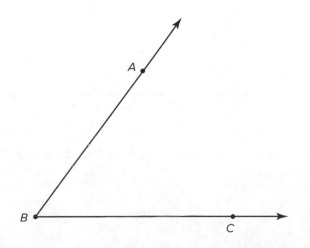

Step 1: Put the point of the compass at point B, which is the vertex of the angle. Make an arc that intersects both sides of the angle. Label the points of intersection D and E.

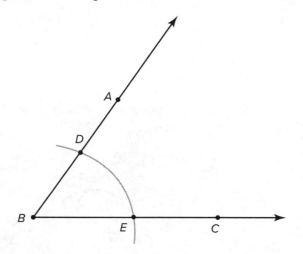

Step 2: Place the point of the compass at point D and make an arc in the interior of the triangle. Place the point of the compass at point E, and make the same sized arc. Label the point where the two points intersect point F.

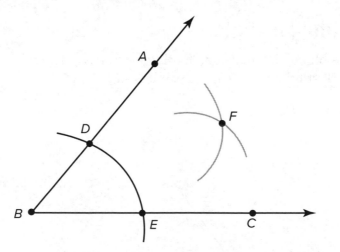

Step 3: Construct a ray connecting point B and point F. $\overrightarrow{BF}$ bisects angle ABC.

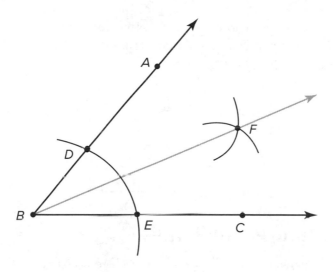

Angle *ABF* = Angle *FBC*

Constructing an Equilateral Triangle

An equilateral triangle has three equal sides. Constructing an equilateral triangle with a straight edge, a compass, and a pencil is painless. Just follow these three steps.

Step 1: Use the straight edge and the pencil to draw a segment, which will be one side of the triangle.
Mark the end points of the segment *A* and *B*.

A •————————————• *B*

Step 2: Measure the length of the segment *AB* using the compass. Put the point of the compass at point *A*. Draw an arc with a radius that is the same length as segment *AB*. Put the compass on point *B*. Draw another arc with a radius that is the same length as segment *AB*. The two arcs should intersect. Label this point of intersection *C*.

A •————————————• *B*

Step 3: Connect point *A* to point *C*. Now connect point *B* to point *C*. The result is an equilateral triangle. All three sides are the same length.

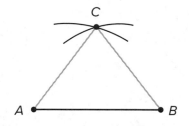

Constructing Parallel Lines

Use a compass, a straight edge, and a pencil to construct a line through a point parallel to a given line. Just follow these four steps.

Construct a line parallel to line *n* through point *P*.

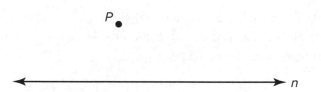

Step 1: Draw a line through point *P* that intersects line *n*. This line is called a transversal. Label the point of intersection of the line *n* and the transversal point *A*.

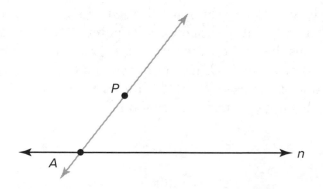

Step 2: Put the point of the compass on point *A* and draw an arc
through both the transversal and line *n*. Label the points of
intersection *B* and *C*. Put the point of the compass on point
P, and draw the same sized arc. Label the point of intersection with the transversal *Q*.

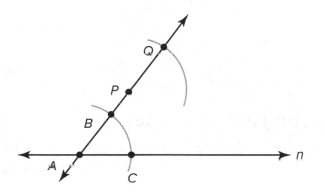

Step 3: Use the compass to measure the distance between points *B*
and *C*. Put the point of your compass at point *Q*, and make
an arc that intersects your original arc. Label the point of
intersection *R*.

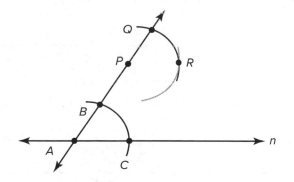

Step 4: Connect point *P* to point *R*. Line *PR* is parallel to line
AC. Angle *BAC* and angle *QPR* are corresponding angles.
When corresponding angles are equal, the lines are parallel.

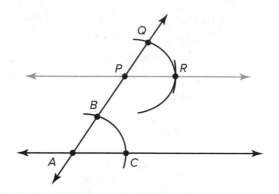

Constructing Perpendicular Lines

Use a compass, a straight edge, and a pencil to construct a line perpendicular to a given line. Just follow these four painless steps.

Construct a line perpendicular to line *m*. Perpendicular lines form right angles when they intersect.

Step 1: Pick a point on the line. Label it *P*.

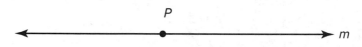

Step 2: Put the point of the compass on point *P*, and draw an arc through line *m*. Label the points where the arc intersects the line *A* and *B*.

Step 3: Put the point of the compass at point *A*, and draw an arc between points *A* and *B*. Next, put the compass at point *B*, and draw the same sized arc. Mark the point where they intersect point *C*.

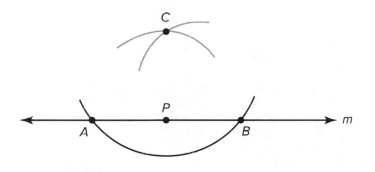

Step 4: Connect point *C* to point *P*. $\overline{CP}$ is perpendicular to $\overline{AB}$. Angles *APC* and *BPC* are both right angles.

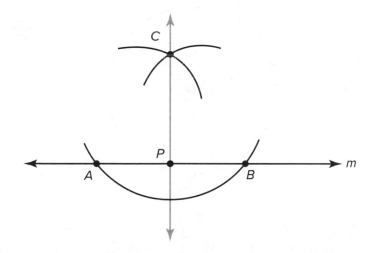

Inscribing a Square in a Circle

When you inscribe a square inside a circle, the four corners of the square intersect the circle. All four sides of the square will be inside the circle. The process is painless. Just follow these four steps.

Step 1: Use a compass to draw a circle. The point of the compass will be the center of the circle.

Step 2: Use a straight edge to draw a diameter of the circle. The diameter of the circle is a straight line that goes through the center of the circle. Mark the points where the diameter intersects the circle *A* and *B*.

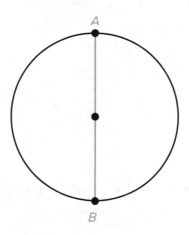

Step 3: Use a compass to bisect segment *AB*. Place the point of the compass on point *A* and draw an arc. Next, place the point of the compass on point *B* and draw a second arc. Draw a line connecting the point where the two arcs intersect through the center of the circle. This line bisects segment *AB*. Label the points where the line intersects the circle *C* and *D*.

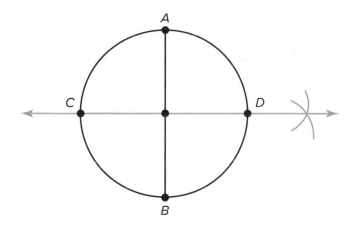

Step 4: Connect points *A*, *B*, *C*, and *D* to create a square

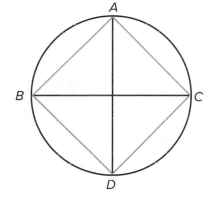

BRAIN TICKLERS   Set # 77

1. Construct a line segment. Construct a segment congruent to it.

2. Construct an angle. Construct an angle congruent to it.

3. Construct an angle. Bisect it.

4. Construct an equilateral triangle.

5. Construct parallel lines.

6. Construct a line perpendicular to another line through a point not on the line.

7. Inscribe a square in a triangle.

(Answers are on page 293.)

SUPER BRAIN TICKLERS

1. Construct a 45-degree angle.

2. Construct parallel lines using alternate interior angles.

3. Construct an isosceles right triangle.

(Answers are on page 293.)

BRAIN TICKLERS—THE ANSWERS

Set # 77, page 291

1–7.   Copy the constructions in Chapter 11 until you can replicate them easily.

Super Brain Ticklers, page 292

1. Hint: Construct perpendicular lines to create four right angles. Bisect one of the angles to create two 45-degree angles.

2. Hint: Alternate interior angles are two angles between the parallel lines on the opposite sides of the transversal. When alternate interior angles are equal, the lines are parallel.

3. Hint: Construct a segment *AB*. Construct two equal segments starting at points *A* and *B*. These segments should be equal to each other but not equal to segment *AB*.

Glossary

Acute Angle: An angle that measures more than 0 degrees and less than 90 degrees.

Adjacent Angles: Two angles that share one side and no interior points.

Alternate Exterior Angles: If two lines are cut by a transversal, the two angles on opposite sides of the transversal, but outside the two lines, are the alternate exterior angles.

Alternate Interior Angles: If two lines are cut by a transversal, the two angles on opposite sides of the transversal, but between the two lines, are the alternate interior angles.

Altitude of a Triangle: The distance between the vertex of a triangle and the opposite side.

Angle: A pair of rays with the same endpoint.

Angle Bisector: A line or ray that divides an angle into two equal angles.

Arc: A portion of a circle.

Area: The number of square units inside a geometric figure.

Central Angle: An angle whose vertex is in the center of a circle and whose sides intersect with the sides of the circle.

Chord: A line segment whose endpoints lie "on" a circle.

Circle: The set of all points equidistant from a given point.

Circumference: The distance around the circle.

Circumscribed Polygon: A polygon whose sides are all tangent to a circle.

Collinear Points: Points that lie on the same line.

Complementary Angles: Two angles whose sum is 90 degrees.

Concave Polygon: A polygon where one or more of its interior angles is greater than 180 degrees.

Conditional Statements: Statements of the form, "If p, then q."

Congruent Angles: Two angles with exactly the same measure.

Congruent Shapes: Figures with exactly the same size and shape.

Convex Polygon: A polygon where all of the interior angles are less than 180 degrees.

Corresponding Angles: If two lines are cut by a transversal, the nonadjacent interior and exterior angles on the same side of the transversal are the corresponding angles.

Cylinder: A three-dimensional shape composed of two parallel congruent circles joined by a straight perpendicular surface.

Decagon: A ten-sided polygon.

Deductive Proofs: Classic two-column proofs that use definitions, theorems, and postulates to prove a new theorem true.

Diagonal: A line segment connecting opposite angles in a quadrilateral.

Diameter of a Circle: A line segment that connects both sides of a circle and goes through the center of a circle.

Dodecagon: A twelve-sided polygon.

Equiangular Triangle: A triangle with three congruent angles.

Equilateral Triangle: A triangle with three equal sides and three equal angles.

Exterior Angle: An angle formed by extending the side of a polygon.

Height: The altitude of a triangle.

Hendecagon: An eleven-sided polygon.

Heptagon: A seven-sided figure.

Hexagon: A six-sided figure.

Hypotenuse: The side of a right triangle that is opposite the right angle.

Inductive Proofs: Proofs that start with examples and lead to a conclusion.

Inscribed Angle: An angle formed by two chords with a common endpoint.

Inscribed Polygon: A polygon with vertices that lie on a circle.

Interior Angles: The angles inside a polygon.

Irregular Polygon: A polygon in which not all the sides are equal.

Isosceles Right Triangle: A right triangle composed of a 90-degree angle and two 45-degree angles.

Isosceles Trapezoid: A trapezoid where both legs are congruent.

Isosceles Triangle: A triangle with two congruent sides and two congruent angles.

Legs: The sides of a right triangle that form a right angle.

Length: The distance from one end of a line segment to the other.

Line: A set of continuous points that extend indefinitely in either direction.

Line Segment: A part of a line with two endpoints.

Major Arc: An arc with a measure greater than 180 degrees.

Midpoint: The single point equidistant from both endpoints of a line segment.

Minor Arc: An arc with a measure less than 180 degrees.

Nonagon: A nine-sided polygon.

Noncollinear Points: Points that do not lie on the same line.

Obtuse Angle: An angle that measures more than 90 degrees and less than 180 degrees.

Octagon: An eight-sided figure.

Opposite Rays: Two rays that have the same endpoint and form a straight line.

Parallel Lines: Two lines in the same plane that do not intersect. Parallel lines have the same slope.

Parallelogram: A quadrilateral with two pairs of parallel sides.

Pentagon: A five-sided polygon.

Perimeter of a Polygon: The sum of the measures of the sides of a polygon.

Perpendicular Lines: Two lines that intersect to form four right angles.

Pi: A Greek symbol that represents the relationship between the diameter of a circle and the circumference of a circle. Pi is written like this: π.

Plane: A flat surface that extends indefinitely in all directions.

Point: A specific place in space.

Point-Slope Form: An equation of a line in the form $y = mx + b$ where m is the slope and b is where the line intercepts the x-axis.

Pentagon: A five-sided figure.

Polygon: A closed figure with three or more sides.

Postulates: Generalizations that cannot be proven true.

Proportion: An equation that sets two ratios equal to each other.

Protractor: A clear plastic semicircle used to measure the size of an angle.

Pythagorean Theorem: A theorem that states that the square of the length of the hypotenuse of a right triangle is equal to the sum of the squares of the lengths of the legs.

Pythagorean Triple: A set of three positive integers that satisfy the Pythagorean equation.

Quadrants: One of four quadrants which are sections of the coordinate plane.

Radius of a Circle: A line segment that connects the center of a circle with a point on the circle.

Ratio: The ratio of two numbers, a and b, is a divided by b. This is written $\frac{a}{b}$ as long as b is not equal to zero.

Ray: A half line. A ray has one endpoint and extends infinitely in the other direction.

Rectangle: A parallelogram with four right angles.

Reflex Angle: An angle with a measure greater than 90 degrees and less than 360 degrees.

Regular Polygon: A polygon with equal sides and equal angles.

Rhombus: A parallelogram with four equal sides.

Right Angle: An angle that measures exactly 90 degrees.

Scalene Triangle: A triangle that has no congruent sides and no congruent angles.

Secant: A line that intersects a circle at two points.

Semicircle: Half of a circle and the connecting diameter.

Similar Triangles: Two triangles are similar if the corresponding angles of one triangle are similar to the corresponding angles of a second triangle.

Slope: The measure of the incline of a line.

Square: A parallelogram with four right angles and four equal sides.

Straight Angle: An angle that measures exactly 180 degrees.

Supplementary Angles: Two angles whose sum is 180 degrees.

Surface Area: The outside surface of a solid shape.

Tangent: A line "on" the exterior of a circle that touches the circle at exactly one point.

Theorems: Generalizations in geometry that can be proven true.

Trapezoid: A quadrilateral with exactly one pair of parallel sides.

Vertical Angles: Opposite angles that are formed when two lines intersect.

Volume: The capacity of a solid figure. The volume of a three-dimensional shape is written in cubic units.

x-**axis:** The horizontal axis on the coordinate plane.

y-**axis:** The vertical axis on the coordinate plane.

Key Formulas

30-60-90-Degree Triangle: The ratio of the length of the sides of a 30-60-90-degree triangle is always $1:\sqrt{3}:2$.

45-45-90-Degree Triangle: Also called an isosceles right triangle, the ratio of the length of the sides is $1:1:\sqrt{2}$.

Area of a Circle: πr^2.

Area of a Parallelogram: base $\times$ height or bh.

Area of a Regular Polygon: $\frac{1}{2}ap$, where a is the apothem and p is the perimeter.

Area of a Square: s^2.

Area of a Rectangle: length $\times$ width or lw.

Area of a Rhombus: $\frac{1}{2}(d_1 \times d_2)$, where d_1 and d_2 are the diagonals of the rhombus.

Area of a Trapezoid: One half the sum of the bases times the altitude.

Area of a Triangle: $\frac{1}{2}(\text{base} \times \text{height})$ or $\frac{1}{2}bh$.

Circumference of a Circle: πd.

Distance Formula: The distance between any two points (x_1, y_1) and (x_2, y_2) is $\sqrt{(x_2 - x_1)^2 + (y_2 - y_1)^2}$.

Exterior Angle of a Regular Polygon: $\frac{360}{n}$ degrees, where n is the number of sides of the polygon.

Exterior Angles of a Triangle: The sum of the exterior angles of a triangle is always 360 degrees.

Interior Angle of a Regular Polygon: $\frac{(n-2)180}{n}$ degrees.

Length of a Median of a Trapezoid: $\frac{1}{2}(b_1 + b_2)$, where b_1 and b_2 are the bases of the trapezoid.

Measure of a Central Angle: A central angle is formed by two intersecting radii. The measure of a central angle is equal to the measure of the intersected arc.

Measure of an Inscribed Angle: An inscribed angle is formed by two intersecting chords with the vertex on the circle. Its measure is equal to half the measure of the intersected arc.

Measure of Angle Formed by Two Intersecting Tangents: Half the difference of the measure of the two intersected arcs.

Measure of Angle Formed by One Tangent and One Secant with the Vertex Outside of the Circle: Half the difference of the measure of the intersected arcs.

Measure of Angle Formed by One Tangent and One Secant with the Vertex on the Circle: The measure of the intersected arc.

Measure of Angle Formed by Two Secants that Intersect Outside the Circle: Half the difference of the measures of the two intersected arcs.

Measure of Angle Formed by Two Secants that Intersect Inside the Circle: Half the sum of the intersected arc of the angle and the intersected arc formed by its vertical angle.

Measure of Angle Formed by a Tangent and a Radius: Always 90 degrees.

Midpoint Formula: The midpoint of a line segment with endpoints (x_1, y_1) and (x_2, y_2) is $\left(\frac{x_1 + x_2}{2}\right), \left(\frac{y_1 + y_2}{2}\right)$.

Perimeter of a Regular Polygon: $n(s)$, where n is the number of sides and s is the length of one of the sides.

Perimeter of a Square: $4s$.

Perimeter of a Trapezoid: $b_1 + b_2 + s_1 + s_2$, where b_1 and b_2 are the bases of the trapezoid and s_1 and s_2 are the sides of the trapezoid.

Point-Slope Formula: $y - y_1 = m(x - x_1)$, where x_1 and y_1 are coordinates of a single point and m is the slope of the line.

Pythagorean Formula: $a^2 + b^2 = c^2$, where a and b are the lengths of the legs of a right triangle and c is the length of the hypotenuse.

Right Triangle Proportions Theorem: If the altitude of a right triangle is drawn, three similar triangles are formed. The sides of these triangles are proportional to each other.

Slope: Given any two points on a line (x_1, y_1) and (x_2, y_2), the slope of the line connecting the two points is $\dfrac{y_2 - y_1}{x_2 - x_1}$.

Slope-Intercept Form: $y = mx + b$.

Sum of the Interior Angles of a Polygon: $(n - 2)180$ degrees.

Surface Area of a Cube: $6s^2$.

Surface Area of a Cylinder: $2\pi r^2 + 2\pi rh$.

Surface Area of a Rectangular Solid: $2lw + 2lh + 2hw$.

Surface Area of a Sphere: $4\pi r^2$.

Volume of a Cone: $\frac{1}{3}\pi r^2 h$.

Volume of a Cube: s^3.

Volume of a Cylinder: $\pi r^2 h$.

Volume of a Rectangular Solid: $l \times w \times h$ or lwh.

Volume of a Sphere: $\frac{4}{3}\pi r^3$.

Index